Seeing the Science in Children's Thinking: Case Studies of Student Inquiry in Physical Science

D1130019

A STAFF DEVELOPER'S GUIDE

David Hammer

Emily van Zee

Education Resource Center
University of Delaware
Newark, DE 19716-2940

HEINEMANN
Portsmouth, NH

T 47773

HEINEMANN
A division of Reed Elsevier Inc.
361 Hanover Street
Portsmouth, NH 03801–3912
www.heinemann.com

Offices and agents throughout the world

This material is based upon work supported by the National Science Foundation under Grant No. ESI-9986846. Any opinions, findings, and conclusions or recommendations expressed in this material are those of the authors and do not necessarily reflect the views of the National Science Foundation.

Library of Congress Cataloging-in-Publication Data
Hammer, David.
  Seeing the science in children's thinking : case studies of student inquiry in physical science / David Hammer, Emily van Zee.
     p. cm.
  Includes bibliographical references.
  ISBN-13: 978-0-325-00948-3
  ISBN-10: 0-325-00948-1
1. Physical science—Study and teaching (Elementary)—Case studies. 2. Physical science—Study and teaching (Middle school)—Case studies.
3. Inquiry-based learning—Case studies. I. Zee, Emily van. II. Title.
  LB1585.H269 2006
  372.3′5—dc22   2006018576

Editor: Robin Najar
Production: Patricia Adams
Typesetter: Argosy Publishing
Cover design: Jenny Jensen Greenleaf
Manufacturing: Jamie Carter

Printed in the United States of America on acid-free paper

10 09 08 07 06 VP 1 2 3 4 5

Contents

Acknowledgments

This book and the video case studies came out of years of collaboration among a team of teachers and project staff well beyond the eight of us whose writing ended up in these pages. The work that went into developing the cases started with the teachers' classes and the research assistants' visits to observe, talk, and of course videotape. That's how we got the snippets of learning and teaching that we then discussed in our meetings—so many meetings—every other week over three school years and for three weeks each summer. We can't do justice to everyone's contributions in the space of these acknowledgments.

Teachers Alison Alevy, Bruce Booher, Cindy Cicmansky, Dana Dreher, Connie Flowers, Charles Gale, Chris Horne, Trisha Kagey, Kathleen Hogan, and Jennifer Peter all worked with us for all three years of the project, and that's what made it all possible: to have a core group of such wonderful, dedicated professionals help shape the project and develop the practices of collecting and talking about snippets. Stacy Catlett-Muhammad joined in the final year and fit right in; Beth Deigan, Kathleen Dunmire, Kathy Jacobs, Dienny Oropeza, and Debi Roberts helped us get the project started.

Loucas Louca was there from the start as a graduate student research assistant (GSRA). For a year he essentially ran the project, and without his energy and abilities, it would never have gotten off the ground. Leslie Atkins, Paul Hutchison, Matty Lau, and Rosemary Russ all joined the project along the way as GSRAs; Laura Lising and David May were post-docs on the project; and Seth Rosenberg was a visiting faculty member. They all shared an understanding and love for both physics and children, and they contributed to the project in many ways, practical and conceptual, from visiting and taping classes, to working with teachers to select and prepare snippets, to editing video, to bringing their insights to project meetings, to offering suggestions on drafts of the manuscript.

To give some particular credits: Loucas taped and did the first round of video editing for Mary Bell's pendulum case, and he did a good deal of work on that and other cases to make the most out of the sound quality. Leslie

taped and edited Kathy Swire's magnets video and did final work on that
and on the pendulum one. Jamie Mikeska taped the first day of her "Falling
Objects" discussion; Paul taped the second day; and Rosemary did the video
editing. Laura taped Pat Roy's discussion about bubbles, and Matty Lau did
the video editing (including our only instance of dubbing!). Seth Rosenberg
taped Jessica Phelan's students talking about the rock cycle. Thanks also to
Bethany Miskelly and Sean Munjal, undergraduates who helped with
transcribing.

All of the staff read and commented on the cases and text of the book;
David May played a particularly valuable role in editing case studies. More
thanks go to Janet Coffey, Andy Elby, and Rachel Scherr for critical feedback
and suggestions, and to Lauren Hammer for her careful reading and comments on the penultimate draft. Benjamin and Joseph Hammer provided
helpful ideas for science fair projects.

We thank Elaine Henry for all of her help throughout, making sure rooms
were scheduled, staff were paid, parking and travel arranged—it's a long list!
Joyce Evans at the National Science Foundation was a wonderful program
officer for our grant—visiting and participating in our meetings, supporting
us at every turn. Carole Stearns gave us helpful advice in the late stages about
where and how to publish. And we thank Robin Najar, Patty Adams,
Elizabeth Tripp, Kevin Carlson, and Pip Clews at Heinemann for all their
wonderful work in production and copyediting.

Last but certainly not least, we are deeply grateful to the students (and
their parents) for giving us their permission to include them in these materials. If you're one of them, seeing your younger self on the DVD and in the
case studies, we think you should be proud.

About the Authors and Contributors

Mary Bell was an elementary school special education and Reading Recovery teacher in Prince George's County, Maryland, for twenty-eight years. Currently, she represents special education as an Instructional Specialist in Curriculum and Instruction at the elementary level.

David Hammer is a Professor of Physics and Curriculum and Instruction at the University of Maryland. He studies the learning and teaching of physics from elementary school through college.

Steve Longenecker was a middle school science teacher for seven years. He now works in the information technology field and spends some of his time as an educational technology consultant.

Jamie Mikeska was an elementary school teacher in Montgomery County, Maryland, for five years. She is now pursuing a doctoral degree at Michigan State University, with a focus on science education and literacy education.

Jessica Phelan has been teaching middle school science in Montgomery County, Maryland, for eight years. She is currently pursuing a second master's degree in school library media.

Patricia Roy taught at a private school in the District of Columbia for four years and has been an elementary school teacher in Prince George's County, Maryland, for fifteen years. She was recently trained as a Reading Recovery teacher.

Kathy Swire is an elementary science and math teacher specialist for Frederick County Public Schools in Frederick, Maryland. She has taught for twenty-one years in grades K–5 in general and special education classrooms.

Emily van Zee is an Associate Professor of Science Education at Oregon State University. She collaborates with teachers in developing case studies of science learning in progress.

CHAPTER 1

A Focus on Children's Inquiry

Imagine You're the Teacher . . .

Suppose you're teaching first grade, trying to get the kids started doing science, and your class is discussing which would fall faster if you dropped it, a sheet of paper or a book. What should you expect to hear in their discussion? What should you hope to accomplish? Or suppose you're teaching third grade and the question is What shape bubble would come out of a square bubble wand? So many questions can come up in elementary school science, from the curriculum, from the children, from things going on in the world: How do rainbows form? Why does the moon change shape? Why do leaves change color in the fall? How do magnets work? What makes day and night? What are clouds? What makes an earthquake? Can a seed grow in sand instead of dirt?

For any topic, in one way or another, you'll need to have some sense of what you'd like your students to get out of exploring it. And you'll need to have some sense, as you watch and listen, of what they *are* getting out of exploring it. It isn't easy to develop that sense of what the students should be and are getting out of their early forays into science. It may seem easy in the abstract—oh, they should see that the book falls faster than the piece of paper because it's heavier—but even that can get tricky: shouldn't they see that heavy things and light things fall at the same rate? (That's Galileo's famous discovery, an idea that gives older students trouble.)

However easy or hard it might seem in the abstract, it's harder in practice, with real students in a real classroom. Jamie Mikeska's first graders talked about their predictions for what would happen if they dropped a book and a sheet of paper at the same time (see "Falling Objects (Day 1)" transcript). As a class, they said the book would fall faster because it had "more strength" than the paper. They said the paper would "float" because "the air [would blow]" it. Then Jamie had them break up into groups to try the experiment.

Now imagine you're Jamie and you've brought the class back together to talk about what the students observed in "Falling Objects (Day 2)".

Teacher: OK, Ebony, why don't you go ahead and begin.

Ebony: To me, first, the paper fell first.

Most of the class reacts in disbelief or amazement ("No way!" "Whoa!" "The book fell first."), but Ebony insists, "No, to me the paper fell first."
Allison helps, saying, "To Ebony, the paper fell first."
Rebecca then speaks up with a different answer.

Rebecca: With him and Julio, twice the book and the paper tied—twice.

Diamond: . . . went at the same time.

Rebecca: They both tied twice.

Diamond: What do you mean, "both at the same time"?

Julio: They both fell down at the same time.

Allison: Same with me, same with me, same with me!

What should you be thinking? Is this progress? Before the students tried it, the class was in consensus that the book would fall first; now they have tried it and they're suddenly all individuals with different ideas. And so far, they don't seem to be converging on the answer that the book falls more quickly!

In another school, Pat Roy's third graders are talking about whether blowing through triangular, square, or rectangular wands will make bubbles in those shapes. Later, they know, they'll get to try it, but before they do, Pat wants them to talk about what they expect will happen. Kedra says that the bubbles will come out in the shape of the wands, and most of the class agrees with her. Is this the moment to go try the experiment? Pat doesn't think so, and soon Louis says he thinks "it will come out in a circle always." Jasmine agrees with him, and so does Zoë.

Zoë says, "I don't think that the rectangle, the square, and the triangle will come out in the same, in that shape, because it can't get all the sides it's supposed to get. If it was blown in that, it would just be, it would be like it is flat."

What does Zoë mean by "it can't get all the sides it's supposed to get" and that it would be "flat"? If you were Pat, what should you think about Zoë's comment? Is it something to pursue in the discussion, or do you think it is enough that she's had a chance to speak and you should give others a turn?

A bit later, Samuel says he's tried it before with the triangle, "and it came out like a circle." Louis says he's done it before, too, which is how he knows the bubbles will be circles. Then Lester says *he's* tried it with a rectangular wand, and "it came out as a rectangle"! Is this progress? Toward what—what should you be trying to accomplish? Has the conversation gone on long enough? Has it gone on too long?

Mary Bell's fifth and sixth graders are debating what would happen to a pendulum (see "The Pendulum Question" transcripts and DVD). Mary has a string with a metal washer on the end. Swinging it back and forth, she'd asked what would happen to the washer if she let go of the string at the highest point of the swing. Chris and Ike had each put an answer on the board—Chris thought the washer would go up and out; Ike though it would go out but not up—and students sided one way or the other. Then Victoria gave her idea that the washer would fall straight down, "because gravity is gonna push it down," and the debate picked up.

> **Brandon:** I disagree with Victoria because, and I kind of agree with Ike, because if the washer, and if you swing it, and it's at the top point, it's gonna go flying up some and then it's going to drop down.

> **Shadawn:** I kind of agree with Amber because, like, it depends on how fast, how fast it's going. Because I think, like, if you, if it's going really fast and you cut it, it's gonna fly somewhere and do all the curves and stuff, but if you, if it's going really slow and you cut it, I think it's just gonna go straight down.

> **Teacher:** Shadawn, can you tell us why you think that? Is there is something that you know of that maybe makes you think that?

> **Mathew:** Can I say something? I agree with, um, Shadawn because it's kind of like, you, you have a little, like you know how sometimes on movies and things and real life like that, they have lakes or swimming pools and you have a little rope and you run and grab onto the rope and then fly and then let go and you go flying over to the side? That's just like that, the washer. It depends on how much force is on it.

What should you think about how Victoria and the other students are doing? Is it time to stop the conversation and have them try the experiment? If you believe like Victoria that the pendulum will fall straight down, in what way will that affect your impression of how well the students are doing at scientific inquiry?

One more: Steve Longenecker is wrestling with his formal responsibility to meet Maryland Science Content Standard 4.8.3, for eighth grade, which states that Students Will Be Able To (SWBAT) "distinguish between chemical and physical changes based on observable properties." The class is considering whether salt dissolving in water is more like an Alka-Seltzer tablet fizzing (a chemical change they have observed) or more like glass beads placed in water (a physical mixing). The textbook's answer is that salt dissolving in water is a physical change, more like the beads in water than the fizzing tablet. But the students haven't read that yet, and right now they think it's the other way around. Alex starts by noting that the salt and tablet both disappear.

> **Alex:** Well, the Alka-Seltzer and the salt actually mixes with the water; the beads can't really . . .

Lincoln: I'm saying that the beads take up a lot more matter than salt or Alka-Seltzer. They take up a lot more matter, a lot more space, volume, than the . . . I'm just saying what Alex is saying.

Teacher: No, I think you're saying something different. You're saying when you add the beads to the water, they take more space up in your container, but when you add the salt or the Alka-Seltzer to the water, it doesn't really change the amount of space that the water is taking up.

Lincoln: Yeah.

Christine: I agree with them two. And also because if you put salt or Alka-Seltzer in, you couldn't easily just put your hand in and take them back out. But with glass beads you can just put your hand in and take them out.

What should you be thinking—about your students' reasoning, about standard 4.8.3, about how to proceed? That phrase "based on observable properties" in the standard raises a problem: What are the observable properties on which students should base the classification of salt dissolving as a physical change? And, of course, 4.8.3 isn't your only concern.

The Challenge of Seeing and Hearing Inquiry

Jamie Mikeska knows her students had it right in the conversation beforehand, when they'd predicted the book would hit the floor first, and she knows that something's amiss about their having observed the paper hitting first or at the same time. Pat Roy knows the students are right who say that bubbles will come out round, no matter the shape of the wand, and she knows Lester can't actually have seen rectangular bubbles. For Mary Bell and Steve Longenecker, too, it may seem easy to assess whether the students have the right answer.

Of course all of these teachers, and the education community in general, have more in mind than right answers. We all want to engage students' interest and participation, for example. These also seem easy to assess. When you watch the videos of these classes, you'll quickly form a sense of who is participating actively, who seems to be listening, and who isn't paying attention. More than that, for the most part, you'll probably be able to talk about what gives you that sense: Signs of participation include children speaking, looking at the person who is speaking, and raising their hands. Squirming, looking off in other directions, vacant facial expressions, and so on are signs of inattentiveness.

Your sense might be wrong, of course. The child who seems fidgety and inattentive is sometimes following every word; the child who speaks up with the right answer might be making a random guess or parroting another child. That's the nature of human judgment: Doctors sometimes misdiagnose symptoms, judges sometimes misinterpret testimony, and teachers sometimes mis-assess students. This isn't a black-and-white issue, not at all. But with respect to some things, including right answers and participation, we all have a relatively easy time talking about what we want, what we expect, and what we see.

With respect to other things, we have a much harder time. If it's sometimes obvious that what a student says is correct—the bubbles will come out round—it's not usually so obvious whether the student understands why that answer is correct. Many of Pat's students are saying bubbles always come out round, but what do they understand as the reason for that? Mary's student Victoria goes to the board and draws her idea that the washer will fall straight down, and if you know the right answer, it's a simple matter to recognize that she's got it. But why does she think that's the answer? All she says, at this point, is "gravity is gonna push it down," but by that reasoning, everything should always fall straight down. Probably there's more to her thinking than she's said, but she hasn't said it. Children's understanding is clearly important, but it's not very easy to assess.

What about *inquiry*, which has been at the center of so much discussion in state and national standards? In Montgomery County, Maryland, the first section of the document that gave Steve standard 4.8.3 is devoted to "skills and processes": *"Students will demonstrate the thinking and acting inherent in the practice of science."* It includes "scientific inquiry" standards, such as 1.8.3, which states that Students Will Be Able To (SWBAT) *"use observations, research, and select appropriate scientific information to form predictions and hypotheses"* at grade 8 (the middle number indicates the grade level), in a progression that starts with 1.3.3 that SWBAT *"use observations and scientific information when forming predictions."* There are also "critical thinking" standards, such as 1.3.13, that SWBAT *"identify that individuals are free to reach different conclusions provided that supporting evidence is used,"* which develops to 1.8.13, that SWBAT *"critique scientific information and identify possible sources of bias."*

Similar sorts of goals appear in just about every set of state and local standards. Everyone agrees they're important, and it's easy to endorse them in the abstract. They are usually not so easy to pursue in specific moments of learning and instruction, however, because student inquiry is seldom clean or clear (let alone correct). Especially at the beginning, what students have to say is often confusing, contradictory, or ambiguous. Were Jamie's students "using observations and scientific information" when they predicted that the book would hit before the paper, or later when some of them observed otherwise? What is "scientific information" in these contexts? The first graders seem accepting of the notion that "individuals are free to reach different conclusions" for which they have evidence; does that mean Jamie should be content to let the students think that "to Ebony, the paper fell first"?

Everything about focusing on student inquiry gets harder when you *get down to cases*, real examples from real classes, where the "best practices" are hardly ever clear. Then things are messy and ambiguous in ways teachers can't escape. It's not possible to control inquiry in the way it's possible to control other things. If we want students to learn to think for themselves, we've got to let them think for themselves, and that means contending with variety and uncertainty. We can anticipate what they might do, but in the end we need to be able to recognize and respond to what they actually do, and as every teacher knows, no two classes are alike.

The wonderful thing is that there's a lot there to see. "The whole of science is nothing more than a refinement of everyday thinking," said Einstein (1936, 349), and children's everyday thinking is filled with the beginnings of science. For educators, it's a matter of learning to recognize them. . . .

- In Jamie's students' reasoning about why the book should fall first—that the book is "stronger" in some sense, and that the paper is more affected by the air: they are thinking about tangible causes and effects, what physicists call *mechanism*. And there are beginnings of science in Rebecca's sense that seeing something happen "twice" is more convincing than seeing it happen once: that's a first-grade version of what scientists call *reproducibility*.

- In Pat's student Zoë's explanation that a bubble can't come out as a rectangle because "it can't get all the sides it's supposed to get": she is also thinking about mechanism, trying to figure out what it is about the formation of a rectangular bubble that she thinks won't work.

- In Mary's student Shadawn's idea that the answer to the pendulum question "depends on how fast it's going": she is identifying a plausible casual factor in the answer. And in Mathew's use of the example of a swing: he is making a reasonable connection to a related, familiar experience.

- In Steve's student Christine's argument that salt dissolving is more like Alka-Seltzer fizzing than like glass beads in water because "you couldn't easily put your hand in and take them back out": she is introducing the idea of *reversibility* into the discussion, a new level of abstraction for thinking about how to classify change.

These aspects of the students' thinking aren't so easy to notice, but they're important. If Mary had noticed only that Shadawn and Mathew still had the wrong answer, she might have discouraged what was actually productive thinking, and the same goes for Steve listening to Christine. If Jamie and Pat had paid attention only to the answers students gave, they might not have supported their students' efforts to explain those answers or develop their sense of mechanism. In the next chapter, we talk more about what might be valuable for science in children's inquiry (other than right answers and active participation); what kinds of things teachers should be glad to see and trying to encourage. We talk about helping students learn to approach science—physical science in particular—as something that builds from their tangible, familiar, everyday sense of the world, about what aspects of that everyday sense are useful, and in what ways.

And then we get down to cases. That's the core of this book: six case studies of children's inquiry, from first grade through eighth, written by their teachers and focused on the children. We've made that our focus rather than our instructional methods, because in our view, finding the best practices at any particular moment in any particular class is a subtle, difficult matter that relies on the teacher's judgment about the students' thinking. What do they need, *these* students at *this* time? What are they saying and doing and thinking? What are the strengths in their reasoning that the teacher should respect, and what are the weaknesses that the teacher might try to address? This is a book about taking on the challenge of understanding children's thinking, its merits and flaws, and the possibilities it affords for building curriculum. It is about interpreting the meaning of what children say and do, in specific instances, and making judgments from there about how to proceed.

How We Developed These Cases

The cases in this volume came out of a collaboration of K–8 teachers and university researchers over three years. We met in three three-week workshops during the summers and in two-hour meetings roughly every other week during the school years. During the summers, we worked on science ourselves, studying topics in physics (motion, light, electricity) both to understand those topics and, as or even more important, to understand what it means to learn science. The teachers' inquiries in science will be the topic of another book. This book presents the teachers' case studies of their students' inquiry, work we did mainly during the school years.

Collecting Snippets of Children's Inquiry

The process of developing these case studies began with collecting snippets of science learning. Mostly these snippets were of conversations, recorded on videotape (sometimes only audiotape) and transcribed for closer study; some consisted of or included samples of students' writings or drawings. We thought of snippets as small samplings of the flood of information teachers take in and interpret all the time. By looking at these samplings, we could slow things down, compare ideas, and generally get practice with interpretation.

The teachers brought snippets into our meetings, and we'd break up into groups of four or five to watch the videotapes, read the transcripts, and look at the materials. The teacher would talk about what had led to the moment the group was examining and give everyone a sense of her or his thinking during the class and afterward. Every case in this book started out as a snippet, and our hope is that you'll use the cases as we did, as opportunities to think about student thinking.

The first challenge we had to face was simply to collect useful snippets of children's inquiry—we needed to be able to see and hear children thinking in science class. We quickly found that's not always possible. Traditional practices of science teaching don't really make room for student thinking, so in order to get snippets of student inquiry, teachers in the project had to experiment with different approaches. This step of making opportunities for students to express and explore their ideas was new enough in itself that it occupied our attention for most of the first year. We had help from Karen Gallas' book *Talking Their Way into Science* (1995), which describes Gallas' technique of having "science talks," in which the children pose the questions and control the conversation. In a Gallas science talk, the teacher doesn't control or provide substance; her role is limited to facilitating and moderating, and even that role fades as students become better at having these talks. For much of our first year, our conversations focused on what teachers can do to bring out student thinking.

As we got better at engaging student thinking and attending to it in class, we found that Gallas science talks were only the beginning, and like Gallas, we moved past the methods she described. We ended up adopting the term *science talks*, but we used it to refer to a broad range of activities that engaged and revealed student thinking, that got students *talking* about their ideas. The teachers spoke of using science talks, but very often that meant conversations around questions posed by the teacher or the curriculum. We found it was usually more productive for the teacher to play a substantive, ongoing role in structuring the talks, in particular by responding to needs the teacher recognized in the students' inquiry. The best approach depended on the particular students at the particular time. Some students in some moments may be best served by the teacher holding back, as Gallas described; at other times it may be better for the teacher to intervene, to draw out a particular strength or to address a particular need.

Our work and these case studies are about recognizing those strengths and needs in children's inquiry. The step of shifting away from traditional approaches was only the first for us, the step that allowed us to begin to think about children's inquiry. To see and hear children's inquiry, teachers need to make opportunities for it to happen in class. Gallas' science talks are one way to start doing that; our science talks comprised many others as well. Teachers can and do draw out student inquiry through project-based learning, science fairs, computer-based interactive environments, structured collaborative groups, and so on. We started with science talks, so you'll find lots of references to them here, but we could have started in other ways. Ultimately what's important to us is the children's inquiry. There are lots of activities, materials, and strategies in science education reform that provide teachers with ideas for getting students engaged and participating; we're not wedded to any of them. *We're not promoting any teaching method other than attending to the children's thinking.*

So these are not case studies of teaching strategies; that's not our intent, and if you read them hoping to find new ideas for methods and curriculum, you might be disappointed. Teacher preparation and professional development programs and publications devote a lot of attention to methods. That is important, no question, but the methods should depend on how the students are thinking and behaving. No method is the best choice all the time; we should focus on the perception and interpretation that goes into choosing how to proceed in particular moments with particular students.

Talking About Children's Thinking Rather than Talking About Methods

At the outset of the project, it was not very long before teachers came in with snippets that showed children thinking. In most classes, we found, it just wasn't very difficult to get the kids engaged. It gets more difficult as they get older, into high school and college, probably because by then they've learned some things we wish they wouldn't have about what happens in science class. In elementary school, though, they can still get excited and involved—for many students it seems like the opportunity they've been waiting for all along, to be able to talk, to have ideas and express them, to have the floor. Kids love to tell us what they think, as long as they know we won't make fun of them or put them down. Convince them it's safe to come out and say what they think, and they will.

Only a month or two into our project, that's what we were seeing—lots of snippets of students participating, with interest and enthusiasm, and having a lot to say. It was exciting and gratifying for all of us so quickly to see results from giving students the time and opportunity to talk about their thinking.

But this success brought a more difficult challenge, the one at the core of the project and this book: It's not very hard to see *that* students are thinking, that they're participating, that they're motivated. Understanding *what* they are thinking is another matter entirely, and it's essential if we're genuinely going to take student reasoning seriously. We spent a fair amount of the first year in this place, delighted to be getting the students thinking, but sitting in groups during meetings not sure what else there was to say about particular snippets.

We've found the same thing in our workshops ever since: It's typically not very easy to get people to talk about children's thinking rather than about the teacher's actions. That's partly because it's often hard to figure out what they're thinking, but then that's something to notice in itself: Are children struggling to articulate their ideas? But the main reason it's hard to focus on children's thinking is simply that we're all most aware of and interested to discuss what *we* do. It's not an accident that courses for teachers in colleges of education are called *methods* courses; we're all naturally inclined to make what we do the main topic of conversation.

So, at the outset of the project and in workshops we've run, conversations would drift from analysis of the students' thinking back to teaching strategies. Those teaching strategies, of course, were often based on impressions of children, but typically about children in general rather than the ones specifically in the snippet. We all have ideas about what kids are like and about what kids really need from school. But the challenges of talking about snippets include focusing on *these* kids in *this moment*. In a sense, we were facing a difficult aspect of scientific inquiry ourselves, the challenge of suspending expectations to consider the evidence at hand. We may know of tendencies we've seen in kids, but is that what we see in this snippet before us?

In some cases, the evidence belied expectations. One important example, not about the substance of the children's thinking but about their attitudes, was the common expectation that children would be upset or frustrated if the teacher didn't tell or show them the right answer to a question. When we read and watched snippets, the concern came up often that the teacher wasn't helping the children get the answer, with the worry about how this would make children feel. When we pressed ourselves to look for evidence in the snippets themselves, in what the children were actually saying and doing, we found it didn't show children fretting about it very much. In the end, we decided grown-ups are the ones who are impatient to get to the right answers! Kids don't seem to be; somewhere along the line they must learn it from us.

In all, we found the second step more challenging than the first. While it was relatively easy to get students started thinking, it was not so easy for us to get started thinking about their thinking, specifically and in detail. We just weren't used to the idea of really listening to what students were saying, for the substance of their thinking rather than simply to see their participation and motivation. Gallas' book and strategy of science talks had helped with the first step. To take the second step, it helped us to structure our conversations in meetings around close readings of the texts of the snippets.

Interpreting Children's Thinking

In science, we all expect, progress often requires careful study of materials and phenomena. You wouldn't be surprised, for example, to open the door of a science class and find students examining different specimens through a magnifying glass, whether rocks or insects or leaves, and comparing and contrasting them in detail. And it would seem perfectly appropriate to find students making careful study of particular phenomena, whether of salt dissolving in water, the spectrum of light from a candle flame, or a ball rolling down a ramp.

We ask educators to do the same sorts of things in trying to understand children's thinking. Our specimens are moments from science class, and if you opened the door to one of our seminars, you might find educators poring carefully over transcripts of a particular discussion, looking closely at a particular line, or playing and replaying a few seconds of video to examine a student's gesture. Or they might be examining a student's drawing or essay or journal entry, arguing about details that a layperson would consider trivial. We think these careful practices of attention to student thinking should make up a substantial part of teacher education and professional development.

In this project it was helpful to get started by taking the transcripts in pieces. We'd watch some segment of the video and then go back over it much more slowly, rereading about ten lines of the transcript, either aloud or while watching that segment of the video. Then we'd go through each line to check how we understood what the student was saying. The group would note various interpretations of meaning, often struggling with particular statements or phrases. In many cases, we found differences of opinion about particular statements we could not resolve. When Ebony said, "To me, the paper fell first," did he mean that when *he* tried it, the paper fell first? Or did he mean that from his perspective, he saw the same event differently from others? What did Shadawn, talking about the pendulum, mean by "do all the curves and stuff" if the pendulum is going really fast? Was she echoing the way Victoria had spoken earlier, or was she referring to Ike's or Chris' answer, which each included curved lines on the blackboard?

Of course, the teacher presenting the snippet knew the students best and had the most information about what was going on. Even so, it was typical for the teacher to leave our discussion with different ideas about the class, from others' suggestions or from having noticed something new on her or his own. It was always up to the teacher to decide what to draw from everyone else's suggestions. Very often the interpretation is uncertain, and that in itself is worth knowing. The evidence in snippets, as in class, is always incomplete, and it's important to be aware of how we may misinterpret students' meanings. A teacher's sense of how a student was thinking, like a doctor's of a patient's well-being or a judge's of a witness' testimony, can be mistaken.

Only after discussing the different possible interpretations did we allow ourselves to talk about the teacher's actions and other possibilities, and we tried to be careful about making judgments. Depending on the interpretation of a moment in class, everyone could have different ideas about what the teacher could do. Even if we agree on an interpretation for what is going on (and in many cases that's asking a lot!), there's always a variety of reasonable ways to respond. That variety puts the teacher who presents a snippet in a terrible position: No matter what the teacher chose to do, someone could always point out another option she didn't choose, another opportunity she didn't take. Jamie could have reminded the students of their prediction that the book would fall first; had a student conduct the experiment in front of everyone; done it herself and asked what everyone saw; and on and on— there are so many possibilities! Any teacher who presents a moment from her class risks criticism for what she should have done, because *there are always other options*. That's especially true of classes that engage children in inquiry: if the class is going well, there are many more opportunities than there is time to pursue.

We tried not to be so naïve as to think someone could say what the teacher *should have* done in any particular situation. At the same time, we felt it was appropriate and important to consider the range of possibilities. That's what makes the interpretations meaningful and relevant, to think about how they might bear on the teacher's actions. We spoke of elaborating a "menu of possibilities," using the metaphor to remind us that we often have to choose from a variety of reasonable options—there are lots of good choices on the menu, but we can't pick all of them. The DVD-ROM includes the prompt sheet we used to structure our conversations about snippets.

This practice of close, careful reading and discussion helped us attend more closely to what children were saying, and it helped us focus on the particularities of the snippet at hand. That's the benefit of case studies, in teacher education as well as in other fields where they're commonplace—art, business, medicine, law. Insights about an individual case don't apply to all cases, any more than insights about a specific work of art apply in general. Of course, it's not possible to stop and analyze each utterance during class. But reading closely in our meetings helps us listen closely in our classes, to be more attentive and perceptive to the substance of student thinking. We become connoisseurs of student thinking in science, so we notice more as it happens.

The Book's Organization

We start, in Chapter 2, with a framework for thinking about the beginnings of scientific inquiry in children. What might we see children say and do that could be a starting point for their "refinement of everyday thinking"? What, in other words, are the productive resources we can identify in children's knowledge and reasoning?

Then we turn to the case studies. You've already had glimpses of some of them. Chapter 3 will give an overview of the materials and some general suggestions for how to use them productively. Chapters 4–9 present the cases. In Chapter 10, we'll talk about moving forward from here, including suggestions for how to focus on children's inquiry in science instruction as well as suggestions for preparing and discussing your own case studies.

There are some additional comments and references to other materials in the Notes section at the back of the book, starting on page 177. These notes are organized by chapter and page number, with phrases from the text used to identify what they concern. (We decided not to include any marks within the text itself based on feedback from teachers who found them distracting.)

CHAPTER 2

The Beginnings of Scientific Reasoning

What sorts of things could be the beginnings of scientific reasoning in children? Not everything they say and do, surely. It's nice to think of children as natural scientists, but aren't they also natural poets and storytellers and jokesters and more?

Our purpose in this chapter is to be more specific about what parts of children's knowledge and abilities might be the beginnings of science. Of course, we can't do that without talking about science. Recall that Einstein said, "The whole of science is nothing more than a refinement of everyday thinking." What everyday thinking could he have had in mind? And how is it refined into science?

There are three major sections to this chapter. The first two, "Common Sense" and "Everyday Reasoning," are about the knowledge and abilities children already have that we want to recognize in the case studies. The third section, "Progress Toward Science," is about what will come of these beginnings later on.

Common Sense

By the time any of us reach school age, we've got all kinds of experience about the physical world. There are differences among children, of course, for reasons of culture, opportunities, and disabilities, but if you stop and think of what any particular child knows, you'll come up with much more than you could ever list.

Try it. Imagine a kindergartener and ask yourself whether she's likely to know about what it's like to lift and drop rocks or books or tissue paper; about how it feels to touch a candle flame or an ice cube; about what water does when she pours or drinks or swims in it. Does she know about seeing things through a window or her reflection in a mirror; about the sounds of hitting drums or guitar strings or just her abdomen; about walking and running, slipping and falling, or slipping and *gliding*? Could she predict what would happen if she were to step on a house of cards—if that's not something she's actually done—or what would happen if she were to pour a jar of ink onto a tablecloth, or if she were to try to use a bowling ball to play soccer?

For any child starting school, the answers to most of these questions, and to so many others you could ask, would be yes. A five-year-old is full of useful knowledge about the physical world, and of course he's learning more all the time. With all his knowledge and experience, there's hardly any question that could come up in physical science that would be entirely new to him. There's almost always going to be something he knows that connects to the topic, whatever it may be. This is the sort of "everyday thinking" Einstein had in mind.

To say just a little more, think about your own everyday knowledge. You know, for example, what it's like to ride in a car or a train or a plane. What do you feel? In particular, think of when it's moving smoothly along, at a steady speed in a straight line. You know that unless it's noisy or bumpy, it doesn't feel much different from when the vehicle is sitting still. You might even fall asleep in your seat, and when you woke up, if you didn't open your eyes, it might be hard to tell how fast you were moving. Or think of riding in an elevator. You know that you feel a little lurch when it starts and again when it stops, but in between it can be hard to tell you're moving at all. (And consider this: someone who'd never experienced riding in a vehicle might think that the pressure of being pushed along that fast would *hurt*.) These examples of everyday knowledge were the beginning of Galileo's and then Einstein's theories of relativity. (Well, Galileo talked about what it feels like to ride on a ship traveling in smooth water, not planes or cars, and Einstein liked trains . . .) So when Einstein wrote "a refinement of everyday thinking," he really did mean *everyday thinking*.

That thinking does get refined, however, and just as refining natural resources can lead to products that look quite different from the raw materials, refining everyday thinking can lead to knowledge that looks quite different from common sense. That could have a lot to do with why people often think of science as contrary to common sense: they haven't followed or taken part in the refining.

Misconceptions

In recent years, a popular view among science educators is that students have lots of misconceptions. The way most people tell it, students arrive at school with lots of wrong ideas that they need to overcome in order to make progress. Worse, they hang on to those wrong ideas pretty tightly. So, the story goes, science teachers really have to make it a mission to ferret out and fix those ideas.

This isn't quite what the research showed, however; in fact, in some ways it's the opposite of what the researchers claimed. They were working from constructivism, the view that people *construct new knowledge using knowledge they already have*. That is, students aren't blank slates waiting for new information. Their knowledge about the world has to be the starting point for any meaningful learning. Research on misconceptions argued that students' prior knowledge is *intelligent* and *useful*. Science teaching, the research said, has to *engage those ideas* or students will dismiss it as irrelevant.

And the research gave evidence that that's what was going on: By ignoring the sensible ideas students had already formed, science teaching was making itself irrelevant to how students thought about the world. Students learned to give certain answers in class, but that had no effect on how they really thought. They still had their earlier ways of thinking—the inherently sensible ideas researchers named misconceptions.

It might have been a poor choice of terms. The researchers used the name to highlight how students' ideas differed from the information science teachers hoped to impart. Some researchers tried to use other words instead, including *preconceptions*, to emphasize that the ideas were productive, constructive steps toward expert understanding.

The important thing to realize, though, is that this research never showed misconceptions or preconceptions to be problems and obstacles to overcome. Unfortunately, most people have misinterpreted the findings, seeing misconceptions as impediments, and in this way misconceptions research has had almost the opposite effect from the researchers' intentions. Instead of raising respect for students' prior understandings, it has convinced many educators that students are *worse* than blank slates; they're slates with wrong ideas written on them in hard-to-erase chalk.

What Are the Productive Resources?

One reason people came to the "obstacles" interpretation is that the research didn't do a good job at explaining how misconceptions are useful steps. Most studies focused on showing that students hold these reasonable but incorrect ideas even after instruction. So it's only natural for teachers and education professors to see misconceptions as obstacles. If all you're thinking about is that the misconception is wrong, then of course you're going to try to get rid of it, not take advantage of it. And you're still left with the question of What's the good stuff? If students *construct new knowledge using knowledge they already have*, what's the stuff they already have that they can use?

Answering this question was difficult in misconceptions research, mainly because it treated each misconception as the *one way* someone had to think about the particular topic.

For example, one famous misconception is that the reason it gets hotter in the summer is because the earth is closer to the sun. It's a reasonable idea! In fact, it fits nicely with pictures of the earth's orbit around the sun that show it getting closer to and farther from the sun over the year.* If we think of it as the one way students have of thinking about seasons, then we have to

---

* Earth's orbit is an ellipse, and it's true that it moves closer to and farther from the sun over the year. But not very much: it's only about a 3 percent difference. And, as it happens, earth is closest to the sun during *winter* in the northern hemisphere.

dissuade them of it to make progress. That would mean *eliciting* the idea—getting students to express it—and then *confronting* the idea with evidence that it is wrong.

That's the usual interpretation: students have in their heads the idea that summer happens because the earth is closer to the sun; it's their one way of thinking about it; and it's pretty firmly engrained.

Another Interpretation: Productive Resources

But there's another way to interpret the misconception about summer. It could be that many students have never given the matter enough thought to have any firmly engrained ideas about it. Asked why it is hotter in the summer, they do a quick search through their common sense and find the idea that *closer means stronger*: the closer you are to the source of something (heat, sound, odor, and so on), the greater its effect. *That's* the idea they have firmly engrained in their common sense—closer means stronger—and the misconception we hear is a result of their using that idea to answer this question.

Thinking this way, the misconception isn't something they already know*; it's something they're thinking up at the moment. In another moment, they might think up something else instead, starting from a different bit of common sense.

Here's an experiment you might try. Make a list of friends you think would be willing to participate (not scientists), and divide the list randomly into two columns. Ask each friend in the first column: "Why is it hotter in the summer than it is in the winter?" The chances are good you'll hear many of them say it's because the earth gets closer to the sun. If you give them paper, some of them may draw a sketch of the earth's orbit around the sun as part of their explanation.

Now do something different with the folks in the second column. For each of them, start out asking this question: "If it's summer in the northern hemisphere, is it also summer in the southern hemisphere?" Many of them will already know it's not, whether from their travels or the news. Let that be the topic of conversation for a little while—ask them to explain how they know; give them a chance to explain why they think that is. And then move on to the original question: Why is it hotter in the summer than it is in the winter?

Chances are this time you won't hear as many of them tell you that the earth gets closer to the sun. Instead, you'll hear more about one *side* of the earth getting closer. All that conversation about how it's winter in Australia and South Africa when it's summer in North America and Europe doesn't fit with the thought that the earth gets closer to the sun: if the earth got closer to the sun, it should be hotter everywhere at the same time. So if they don't

---

* When we say *know*, we mean "have in their heads." Using the word this way, knowledge isn't necessarily true. That's what *knowledge* generally means in cognitive psychology and education research.

already have that idea in their heads, they'll be less likely to produce it in answering the question.

In other words, when they do their quick search through their common sense for an idea, if they find *closer means stronger*, they won't be able to use it the same way, because you've made sure they've already found another part of their knowledge. Of course, the people listed in your first column were just as likely to know that summer in the north means winter in the south, but that bit of their knowledge might not have occurred to them at the moment.

You could try other versions of this experiment, too. Spend the first part of the conversation talking about how days are longer in the summer than in the winter. Then when you ask why it's hotter in the summer, it's likely you'll hear more answers about how the sun has more time to warm things up. It all depends on what parts of people's common sense occur to them at the moment they're thinking about the question.

The Many Parts of Common Sense

We don't have just one way of thinking about any topic. We have a variety of ways, which we use variously depending on the circumstances. When someone says that the earth gets closer to the sun, that's not the only way he has of thinking about seasons. He has other ways, too, which he'd use in other moments depending on the situation.

It's easy to find more examples of this. Ask adults, "When can you see the moon?" and most of them will say, "At night," maybe with a quizzical look over why you're asking such an obvious question. They'll do that despite the fact that over their lives they've very often seen the moon in the daytime and haven't thought it was strange. Ask children in the morning if the shadow of a telephone pole will be in the same place in the afternoon, and some will say yes, even though they've seen it in a different place every day after school. Ask people if a glove keeps a hand warm or cool, and almost everyone will say warm, forgetting the parts of their common sense that have to do with oven mitts.

This doesn't happen just in science, of course. Is it OK to tell a lie? You might want to say no, asked that question in the abstract, but if you stop and think about it, you can come up with situations in which it's right to tell a lie. It's rare to find someone who treats "Lying is wrong" as a hard-and-fast law that holds in all situations.

That's not how common sense works, in general, with fixed rules and categories. Common sense—everyday thinking—is flexible and multifaceted, variable and adaptive to the particular circumstances. Pollsters and politicians know this very well—they can influence how we think about a topic by connecting it to different parts of our common sense.

So when you're thinking about common sense, don't picture a mental encyclopedia we consult for answers. Picture instead a rich, varied collection of ideas and experiences and different ways of thinking, knowledge in all sorts of shapes and sizes, many different parts that do many different things.

We call those parts *resources* to help us see them that way: every part is useful in some contexts and to some purposes but not others. Different moments turn on different resources, which enables us to manage the myriad of circumstances we encounter every day. The collection as a whole doesn't fit together the way an encyclopedia does, with a clear organization and a particular bit of information for each topic. It's much more complicated than that, more idiosyncratic and redundant and inconsistent. Thinking about any topic at any moment, you might use any of a variety of assorted resources.

In other words, it's a mistake to think of everyday thinking as a single thing, that "what someone thinks about such and such" generally has a single answer. Rather, "what someone thinks" can have lots of answers depending on which parts of common sense she uses. What someone thinks about why it's hotter in the summer, or whether it's OK to tell a lie, or just about anything else, depends on how the topic comes up.

An analogy for common sense is an extensive toolbox, with lots of different kinds of tools for doing all sorts of things. They're all useful, but for any particular job you might grab the wrong one. If you tried to use a can opener on a bottle, you'd find it wouldn't work so well. But you shouldn't throw the can opener away! The idea that *closer means stronger* doesn't work in the way people tend to use it when they're asked why it's hotter in the summer, but it certainly does work in lots of other situations. It's a resource for thinking that's available in our minds, useful in some situations and not useful in others.

With this view of common sense, *thinking* involves looking through that collection, grabbing different pieces, and trying them out. Faced with some question, the thinker looks through her knowledge and experience and finds resources. When you ask people why it's hotter in the summer, they look through their common sense, and one of the first things they find in there is *closer means stronger*. They might then stop looking, since they've found something that seems to do the job, and so they'll be less likely to find other pieces in the collection.

If they keep looking, there are lots of other pieces they could find, such as that winter in the northern hemisphere means summer in the south, that summer days last longer, or that the sun is more directly overhead in the summer than in the winter. They might also find knowledge about how the sun doesn't feel as warm early or late in the day when it's low in the sky; how clouds or fog can block the sun and make it feel cooler; or how when you're tanning, the side of your body facing the sun feels warmer.

We talk more about refinement in the next section ("Everyday Reasoning"). First, though, we need to talk more about which parts of common sense are the most useful for science.

Different Kinds of Explanations

The list we just gave of what people might find in their common sense for thinking about why there are seasons is very partial. In particular, it is selective for the sorts of ideas that are more likely to be useful. But there are many other resources people might try to use. Ask children instead of your adult friends why it gets hotter in the summer, and you'll probably hear a wider variety of answers. Some of their ideas will be the same sort of thing you'd hear from adults, but some will be quite different:

Because the ground gets hot from volcanoes.

Because people turn on the heaters.

So everybody can go outside and play.

So the plants can grow.

Because it's *summer!*

Of course some children (and too many adults) will answer not so much with an idea as with terminology, if they've gotten the sense that questions in science require answers of fancy vocabulary. So they might answer, "It's the *energy*!" or "It's the *axis*!" or "Because of *solstice*!" without having a tangible sense of what any of those words mean.

That children will say all these different kinds of things in answer to the question gives us a better appreciation of what's *wonderful* about the idea that the earth gets closer to the sun.

First, it's the kind of explanation we want to hear in science, an explanation of physical cause and effect, what scientists call *mechanism*. It's a tangible, clear idea that everyone can picture: If the earth were closer to the sun, it would be hotter. We know it from lots of other experiences, about what it feels like to bring your hand closer to a candle, or to bring the garlic close to your nose, and the like. Even better, that explanation is consistent with other things we know, including about how summer comes once every year and, for those who've come across the pictures, about how orbits have elliptical (oval) shapes. From the children's perspective, that explanation might make great sense.

These are strengths in the explanation we should not take for granted, and if a second grader gave this explanation in class, we should be genuinely happy to hear it. It's a much better explanation than the others.

What makes the others less productive as science? The idea of volcanoes making it hot is also a tangible, clear mechanism: everyone can imagine volcanoes nearby. But it's just not plausible; we don't see or hear any volcanoes erupting around us in the summers! Similarly, the explanation that people turn on their heaters gives a clear mechanism, but children should see the same sort of problem with it. The explanations that it's summer so that kids can go out and play and that plants need it to be warm, on the other hand, don't say anything about mechanism at all. They give reasons, but they're reasons of *purpose*, not of *cause*. They're the wrong kind of explanations for science.

It's important to recognize that none of these responses is inherently unreasonable. Any of them could be perfectly appropriate for some kind of activity. If the activity were to "use your imagination and see how many different things you can think of that could make it hot outside," then the ideas about volcanoes and heaters would be great. If the activity were a discussion about mythology, we'd expect and want reasons of purpose. Adults could happily talk about how summer is warmer than winter because Demeter lets things grow only when she's with her daughter Persephone, who ate some pomegranate and because of that has to spend part of every year in the underworld.

These other sorts of activities aren't science, but there's no reason to expect young children to have clearly sorted out different kinds of conversations. How would they already know science is different from fanciful make-believe and mythology? For most adults, the question Why is it hotter in the summer? automatically triggers a kind of thinking, the sort of plausible, mechanistic cause-and-effect thinking that generates the answer "because the earth is closer to the sun." Part of what children need to learn is to make implicit choices for when to do what kind of thinking.*

Common Sense About Mechanisms

Fortunately, children have a lot of common sense about mechanisms. We've talked about *closer means stronger*, the sense of mechanism children have for understanding that sitting closer to a stereo speaker makes it sound louder, or sitting closer to the fire makes it feel warmer. They've also got a sense of *more direct means stronger*—so turning to face the speaker also makes it sound louder, and it's the side of your body facing the fire that gets the hottest. (At some point we'd want them to start using that sense instead of *closer means stronger* for thinking about why it's hotter in the summer.)

And there's much more. They have a sense of how weight causes falling and of how supporting can prevent falling. More weight causes faster falling (a book falls faster than a piece of paper) and needs more support to prevent falling. Motion can cause an impact, and something moving more quickly will cause a greater impact. Hot things cause burns, and the hotter they are, the worse they burn. Children know about various mechanisms of pushing (blowing on something, pushing it with your hand, pushing it with a spring) and various mechanisms of pulling (with a rope, by sucking, by stickiness). Just think of what physical phenomena children encounter, and imagine the variety of resources they have for answering What makes that happen?

* In fact, the history of science shows that for many years, grown-ups had trouble sorting out which kinds of thinking were appropriate. Long ago, scholars considered it appropriate to give explanations for physical phenomena based on intangible "virtues," as those scholars called them. For example, someone might explain that a stone falls because its natural place is at the center of the earth, and smoke rises because its natural place is in the air: each is just acting out its nature. In the late sixteenth century, scholars began to shun explanations based on such "occult qualities." They argued that explanations shouldn't appeal to indefinable magic; they should describe tangible mechanisms—for many scholars, this was the beginning of *science* as a domain.

That's the starting point for science: children's sense of mechanisms in the world, their basic familiarity with phenomena and their causes. Listening to them reason, we can listen first for whether they are using these resources or others.

Don't forget, we haven't begun to talk about refinement. Common sense of mechanism isn't correct as it is. But it is *useful*, and a child who says it's summer because the earth is closer to the sun is starting to do science in a way that a child who says "so plants can grow" is not. And don't forget either that we're not talking right now about how to teach science; we're talking only about how to recognize its beginnings in students. Whatever the approach or curriculum, whatever your strategies, it's important to have a sense of what you're hoping to see. Watching and reading the case studies, you can look for whether, where, and how children are drawing on their common sense of mechanisms.

Everyday Reasoning

So far we've described science as starting from common sense about phenomena and mechanisms. We've described common sense as a rich, diverse collection of resources—a positively vast collection.

We've also started to talk about how thinking involves picking and sorting through that collection: Ask someone a question, and he'll look through his resources. Sometimes he'll find one quickly and give you a quick answer. Sometimes he won't find anything obvious, and he'll take more time. And sometimes he'll find different parts of his common sense that seem like they should work but don't agree with each other. ("It makes lots of sense that summer would happen because the earth is closer to the sun, but then shouldn't it be hotter everywhere at the same time?") Maybe he'll do something to try to reconcile that inconsistency.

In this section we're going to talk more about these things that people *do* as part of everyday thinking: *shopping for ideas* and *reconciling inconsistencies*.

Shopping for Ideas

Imagine you're working on a crossword puzzle, and you know the answer to 4 across is *that actress who was in that movie*, but you can't think of her name. What do you do? You might start going through the alphabet, trying out each letter until one rings a bell, as people say. You get to *K*, and that does it: Katharine Hepburn. Or imagine you've left your cell phone somewhere, and you're trying to figure out where it might be. Now you go through your day, thinking about all the places you've been, because maybe that will jog your memory. Or, instead of locations, you might think through phone conversations, and to do that you might think through people: "Let's see, who called me yesterday?" Thinking of the person might help you think of the call, which might help you think of the place.

If you're a parent, you've probably taught this sort of strategy to your children:

"Mom, I can't find my teddy bear."

"OK, let's think where it might be . . ."

We call it *shopping for ideas* because you're browsing in your mind for knowledge the way you might browse in a mall for clothing, and the metaphor works pretty well. Good shoppers don't buy the first thing they see. Nor do they buy everything that looks good on the rack—they take it out, try it on, see how it looks. They know that the best stuff might not be in obvious places; they go down aisles other people might miss. And they know how to decide when they've explored enough.

In science, the shopping is for knowledge and experience about the natural world. Studying some question, trying to understand a particular phenomenon, a scientist browses through her knowledge and experience of other phenomena that might or might not be related. "What if I think about it this way?" she asks herself. We talked about all the various bits and pieces of knowledge someone could use to help her think about the seasons, but she could use them only if she dug around in her mind enough to find them. That's how it is throughout science: the ideas are very often there, but in a different section of your mind than you first think to look, like shoes hidden on the bottom shelf in the corner of the store. Part of becoming a scientist is becoming a perseverant and sophisticated idea shopper. It is an important kind of creativity, and breakthroughs in science come from finding new connections to knowledge that already exists.

In the study of electricity, for example, key progress came from scientists finding useful ideas in their knowledge about liquids, about gasses, and even about gears. Students learning about electricity need, similarly, to find useful resources in their common sense, resources they use all the time for understanding things that might seem much too ordinary to be relevant—about hallways crowded with people, or cars on roads, or hoses full of water. No one comes to understand electricity without using ideas he originally developed for other purposes entirely.

This is something to recognize and support in children, this sort of imagination and creativity, their willingness and ability to search through their minds for ideas. They know an awful lot about causal mechanisms; we should be happy to see them poking around through that knowledge. And they often do. We saw lots of examples in the project; here are some from cases that aren't in the book:

■ Fifth graders thinking about how solar water heaters got so hot in their cold classroom considered mechanisms of how heat might accumulate, travel, be blocked, bounce, or "drill" through material; they played with analogies to what they knew about swimming in cold water and eating ice cream, about sunlight reflecting from the moon, and about steam rising from boiling water.

- First graders thinking about whether a seed could grow in sand considered various mechanisms of sand soaking up and holding water like a sponge, either keeping the seed wet or pulling the water away from the seed, and of sand letting water flow easily to get to the seed or flowing too easily and escaping.

- Third graders thinking about what causes earthquakes came up with ideas about how "disturbances" of various kinds could be the cause, including how it might be lava pushing up on the ground from underneath; one student suggested the lava pushes upward the way adding an ice cube pushes up the level of water in a glass.

- Fourth graders trying to explain how a lightbulb gives off light talked about the filament "getting so hot it's on fire a little," being "a really small and controlled fire," and being so "skinny" that it "shows more electricity" than a regular wire, thinking of connections to other experiences, from lightning to explosions to "crystal rocks" that make sparks when you hit them.

Most of these connections the children explore are not correct, and so people who are inclined to assess children's thinking only for correctness might overlook them. The point here is that the shopping itself is important and valuable, even when children don't arrive at the answer. That's actually much more typical of what you'd see scientists doing, day to day in their research, if you had the chance to watch them: looking for answers but not finding them just yet. Important progress happens when someone thinks of a connection that isn't obvious, and it works, but that means good scientists try out lots of connections that don't turn out to work. So we should learn to recognize children's shopping for ideas and trying out possibilities as valuable, including when the possibilities don't turn out to work. Watching and reading the case studies, you can look for students shopping for ideas.

Reconciling Inconsistencies: Arguments and Counterarguments

To review, we've talked about how common sense is an extensive, diverse collection of resources from their knowledge and experience that people use for thinking about the world. Science begins from common sense about physical causes and effects, what scientists call mechanisms. And reasoning in science involves searching through those resources, as people may do in everyday thinking, to become aware of resources that might be relevant to whatever they're trying to understand. The next thing to consider is what happens when multiple resources disagree.

As we've seen, all those various parts of common sense don't say the same things. That's what makes it so wonderfully flexible and adaptive, why it works so well for us across the wide variety of situations we encounter every day. It lets us think differently in different situations, something we don't typically notice. But sometimes we do notice, and then we might find an inconsistency.

For an example outside of science, many people have been wrestling recently with the question of whether it's OK to make copies of music CDs to give to friends. One part of common sense says it's *stealing*: whoever keeps the copy gets something for free that's supposed to cost her money, money that would have gone to the people who made the CD. Another part of common sense says it's *sharing*: someone owns a CD and wants to share it with a friend, just as people used to make copies of albums onto cassettes. The two parts of common sense are inconsistent: one says copying music is bad, and the other says it's good. How do people reconcile this inconsistency?

Some try to justify it one way or the other. If the conflict is between their understanding of what it means to steal someone's property and their understanding of what it means to share their own property, they may decide the first is right and so they need to explain what's wrong about the second. ("The *CD* is my property, but the *music* on it really isn't. All I paid for is one CD's worth of access to the music, so I can't make another.") Someone else may decide to reconcile the contradiction the other way, thinking he's entitled to share the CD and so he needs to explain why that isn't really stealing. ("It isn't really stealing, because I'm not taking anything away from anyone. I'm not removing a CD from a store; I'm making the copy myself.") In this way, reconciling the inconsistency means refining their understanding, of either stealing or sharing or both.

Of course, some people just shrug it off: "Well, yeah, but this is different," they might say, without bothering to explain what's different about it, or "I don't know, I guess it's probably wrong, but I'm still going to do it." That's a choice to accept the inconsistency and leave it alone. That might not be a good choice, for this question, but it's clear that for many questions it's perfectly appropriate to accept inconsistencies. Most of us wouldn't worry about inconsistencies in mythology—they're just stories. We don't expect people to be consistent about tastes. If in some situations you enjoy coffee and in others you don't, that's not something to dwell on—tastes can be fickle. And, of course, people might not be consistent about being consistent! While shooting the breeze on some topic with your buddies in a diner, inconsistencies wouldn't matter in the same way they would if that topic came up in an academic seminar.

But here's the thing: In science, the ultimate goal is to be consistent. We might not succeed, of course, and at any moment there will be unresolved questions, but in the end we want all the ideas and evidence to agree with each other. If you have one idea (e.g., it's summer because the earth is closer to the sun) that doesn't fit with other knowledge or evidence (it's winter on the other side of the earth), you don't just shrug it off. You try to account for it, one way or another, which generally means refining one or both of the ideas. The "refinement of everyday thinking" is toward consistency.

And children do this, too—not always, just as they don't always shop thoroughly for ideas or use their sense of mechanisms, but like adults they have it in them to pay attention to consistency. Educators need to recognize and appreciate when they do so. We've seen lots of examples in our project:

- Third graders were doing an experiment to compare how quickly water cooled in a container that was short and wide versus one that was tall and thin when they noticed a discrepancy: It felt warmer when they put their hands over the short, wide container, but the thermometer in that one measured a slightly lower temperature. Maybe the sun was affecting the thermometers? Maybe the wide container felt warmer because it was bigger? Several of the students couldn't let it go—they needed to reconcile the inconsistency and persisted in trying, even when the teacher needed to end the discussion.
- Fifth graders thinking about solar ovens noticed an inconsistency in their experiences with light. Light from a flame or the sun feels warm, and so it seems light carries heat. But light from the moon or some kinds of bulbs doesn't feel warm at all. They wanted a clear answer: Does light carry heat or doesn't it? And they had ideas: Maybe there are two kinds of light? Maybe moonlight just has too little heat to feel?
- A child in a pre-K class explained that day and night happen because the earth is turning, but that explanation bothered other children because it doesn't feel like the earth is turning. The first child had a way to reconcile the contradiction: the earth is turning too slowly for us to notice, an explanation he demonstrated with a globe.
- A third grader had heard the explanation that rainbows happen because raindrops make light from the sun split into all the colors. But there was something about that she couldn't understand: How would all the colors from all the raindrops end up forming an arch of color in the sky? If every raindrop split the colors from sunlight, why wouldn't the sky look like a big "mixture of colors"?

It's important to recognize that what these children were doing wasn't simply a matter of weighing one set of reasons against another set to decide which was more convincing. That is, they weren't just trying to find the right answer. When they could see a reason on the other side, they didn't just dismiss it, they tried to account for it. Watching and reading the case studies, you'll find more examples of children noticing and trying to reconcile inconsistencies.

The Goal of Coherent Understanding

Striving for consistency is another respect in which scientific reasoning starts from some ways we think every day, but only some—not all the thinking we do every day is a starting place for science.

In some contexts of decision making, the point is to arrive at an answer. If it's a momentous decision, say whether to accept an offer of a teaching position, you might list the pros and cons very carefully. On the side in favor, you might have reasons such as "the salary's good" and "the administration is supportive." On the side against, you might have "it's a long commute" and "the class sizes are too large." Then you'd make a decision based on which side outweighed the other. Having made your choice, it wouldn't be of much value to think any more about the arguments on the other side: sure, the class sizes are larger than you'd want, but it's worth it to have a supportive principal.

That decision-making strategy isn't generally appropriate in science. It does lead to an answer, but it leaves unreconciled objections. In science, when you think you know the answer, you need to go through all the competing arguments and try to *explain why they don't work*. If deciding whether to take a job were a scientific question (it isn't), then if you decided in favor, you'd need to explain what was wrong with your concern that the class sizes are large: You'd need to explain somehow why that concern was either mistaken or not actually relevant. You wouldn't be able to say, "It's a lot of students, that's a valid point, but I'll just have to compromise"; you'd have to figure out why "it's a lot of students" wasn't a valid point in that decision.

This difference reflects how the ultimate goal of science isn't to arrive at the correct answer to a particular question. It is to construct a coherent understanding. That's part of what we should recognize in children.

Those third graders who were thinking about how one container could feel warmer but measure cooler could have simply picked one side to believe and dismissed the other—"really it's cooler; what you feel is wrong." At some moments, of course, that's what they'd choose to do, but in this instance they behaved like scientists. Instead of just guessing one way or the other and ending with that, they picked one side and then tried to reconcile the other side. In fact, they tried it both ways, first supposing the water in the wide container really was warmer and trying to explain why the thermometer was off, and then supposing it really was cooler and trying to explain why it felt warmer.*

Foothold Ideas

The way to make progress with a question, when you're not sure what to believe, is to try believing something and then see where it gets you. If you're unsure whether the answer is *A* or *B*, then try supposing *A*. Assume *A* is true, think about what that would imply, and try to explain what's wrong with the reasoning for *B*. Do that for a while, and if you're still unsure, then try it the other way: suppose *B* is true and try to figure out what's wrong with the reasoning for *A*.

---

* The scientist's answer, by the way, is that these facts are consistent with each other. The core idea is that the wider container loses heat more quickly. So it feels warmer because there's more heat coming away from it. And its temperature is lower because it has lost more heat. (Here's some help shopping, if that doesn't make sense. Think of a good thermos for your coffee. If it's a very good thermos, it won't lose very much heat, and the outside of it will feel cool. If it's not such a good thermos, it will lose heat quickly, and the outside will feel warm.)

We call an idea you choose to assume is true, at least for a moment, a *foothold idea*. Watching and listening to children, we should notice moments when they're behaving this way. It should be part of our assessment of how they're doing. That's especially important when they've picked a foothold idea we know is incorrect, because if we don't recognize the value of their picking a foothold at all, we might not support them in that very scientific behavior.

At some point we want to see them getting right answers, but it's a matter of judgment when that point comes. Along the way, though, they should experience and understand the importance of seeking consistency, because *that's how right answers come to be right*: The "truth" in science is what's consistent. Answers are never right one at a time; they're right in sets connected by reasoning. What we don't want is to teach children it's OK to dismiss ways of thinking without accounting for them, without explaining what exactly is wrong with those ways of thinking.

Science is all about connections among ideas. Knowing an answer without understanding its connections to other ideas and evidence isn't scientific. So we need to notice and appreciate that kind of connected reasoning in children. We want to see them learning to make suppositions—to try out footholds—and to think about implications. We want to see them caring about coherence, and, surely, we want to see them willing and unafraid to think about an idea they're not yet positive is correct.

Progress Toward Science

Scientific inquiry, in sum, is *the pursuit of mechanistic, coherent accounts of natural phenomena*. That's what we want to recognize and support: when children begin that pursuit. We've talked about their having extensive common sense about physical phenomena and causal mechanisms, the ways they might search through that common sense, and the ways they might try to reconcile inconsistencies. These are all aspects of children's knowledge and abilities that are the beginnings of scientific thinking. They won't do these things all the time, so it's all the more important that we notice and appreciate when they are doing these things. They need to learn this is what science is about.

We haven't said very much about other aspects of children's knowledge and abilities that are not as specific to science. In particular, their abilities to communicate are essential, both to express their thinking clearly and precisely, verbally, in pictures, and in writing, and to attend to what others are trying to express. When listening to children's discussions, or reading their work, that is often what we should notice and think about how to address in instruction. We don't see these abilities as any less important than their abilities and inclinations to reason about mechanisms or to reconcile inconsistencies.

These beginnings are the focus of the case studies in this collection. Before moving on to the cases, we'd like to talk about how these aspects of children's everyday thinking underlie the sorts of practices and knowledge people traditionally associate with science. How do these early resources lead toward established scientific principles, such as Newton's laws of motion or the conservation of energy? How do they connect to experimental practices such as posing hypotheses, controlling variables, and collecting and recording data?

So the rest of this chapter is about what comes later, to give a sense for readers who want it of how a focus on children's inquiry will eventually support their developing the knowledge and abilities everyone associates with "real science." We'll talk about how the refinement of everyday thinking leads to *scientific principles, technical vocabulary, mathematics, controlling variables,* and *expert understanding.*

From Footholds to Principles

We just talked about the value of foothold ideas: When you're not sure what to believe, try assuming something is true. Suppose one idea, and then see if you can align the rest of your thinking to agree with that.

We all do that in little ways, as part of everyday thinking. If you're not sure how to arrange the furniture in a room, you might say, "Let's try putting the couch over there," and then you would find out what that would mean for the rest of the furniture. Science builds on that everyday strategy by keeping systematic track of those choices, not just for a moment, thinking about a particular question, but from question to question and day to day. That doesn't mean we can't choose differently; it only means we need to be deliberate about it. We're trying to be consistent in particular moments, but we're also trying to build a stable understanding.

As we set footholds and use them, we end up giving some of them higher priority than others, either because we've used them so much and they've proven to be successful or because it's just too hard to think of how they could be wrong. When there's a foothold we're pretty committed to accepting, then we'll work hard to reconcile other ideas and even evidence to fit with it.

For example, the idea that the earth was the stationary center of the universe was a great foothold idea for a long time. It fit with experience, and evidence, and reasoning: no one could feel the earth moving; everyone could see the sun and the stars and the planets moving. When there was some new evidence and reasoning, from observations that the planets sometimes seem to move backward, it conflicted with the simple model. So scientists tried to reconcile the inconsistencies with ideas about epicycles of motion in the planets: the planets don't just move in circles around the earth, they move in circles around the circles . . . it got complicated. To scientists today, the idea seems silly.

It's very important to realize, though, that it wasn't silly from scientists' perspective then. It was perfectly appropriate for them to try to reconcile other ideas to fit with something that worked so well and made so much sense—that the earth was stationary. In fact, it would have been silly for them to just abandon that idea without a struggle. That remains the case today: Scientists don't abandon ideas easily that have a history of working well for them, and they shouldn't. If an idea makes sense and fits with many aspects of knowledge and experience, we shouldn't be quick to abandon it, even when there's evidence against it. First we should try to explain that evidence, account for it in ways that let us still believe the idea.

In the case of the orbits, explaining the objections got harder and harder, and scientists decided to try a different foothold: the earth orbits the sun. Then the question was whether they could reconcile all the arguments and evidence that the earth was stationary. If we take it as a foothold that the earth is moving, how do we account for the fact that we don't feel that motion? The answer came from a connection to another part of common sense, the one we mentioned near the start of this chapter: if you're riding in a ship on smooth seas (or an airplane in smooth air), and you don't look outside, you can't tell you're moving. Galileo made that argument: the earth is moving smoothly through space, and we're all moving along with it, and we can't tell that we're moving. Over time, the idea that the earth orbits the sun worked; all the other arguments and evidence either fit easily or could be adjusted to fit. Its implications led to new reasons to continue using it as a foothold—and all the reasons to disbelieve it could be reconciled.*

In this way science makes progress toward coherence—toward systems of ideas that fit together and support each other. The ideas that the earth orbits the sun and that we can't feel smooth, uniform motion fit together very well, along with other ideas, and as these ideas were successful in making sense of the world scientists became more and more sure of them. That's what happens: Ideas we choose to believe work, and we become more and more committed to those choices. So we work hard to reconcile other ideas to agree with those commitments. Not all ideas and evidence come into place easily, but they can be reconciled—we can adjust our understanding of them so that they work with the system.

Eventually we are so committed to some foothold ideas that we call them *scientific principles* or *laws of nature*, and we accept them as true. In science, the truth is what fits into a coherent system of thought and experience. That means the truth in science doesn't come one fact at a time: any one idea can be true only as part of a system of other ideas. And so science is the refinement of everyday thinking, from adaptive, flexible, context-sensitive common sense to principled coherence.

* You've got some reasons to disbelieve it in your experience: It takes a continuous push to keep things moving, for example. If you put this book on a table and pushed it, you wouldn't expect it to keep moving if you stopped pushing it. And to get it moving *fast*, you'd have to push it pretty hard! How does this part of your common sense fit with that other part of your common sense about how you can be moving very fast in an airplane but not feel any push on you at all? That's something to work on in a physics course.

The process of refinement is similar to how governments design laws for society. (In fact, the idea that there could be laws to govern society came before the idea that there could be laws that govern nature.) Laws are the ideas we choose to live by, and the role of the courts is to decide what is consistent with those laws. We look for the ideas that work the best as footholds to make our reasoning and decision making as fair and consistent as possible.

As in science, our sense of the best foothold ideas can change as we get new information or people come up with new lines of reasoning. It used to be, for example, that the laws defined owning land as also owning the space above that land, all the way to the heavens: a bird flying over your land was in your property. The invention of the airplane raised a problem: The foothold idea that was the legal definition of property implied that pilots or airlines needed permission from everyone along their route to fly through their property. So air travel would essentially be impossible, not only for all the work it would take to get all the permissions but also because some people would refuse. The idea that air travel should be possible seemed to deserve a higher commitment than the existing definition of property, so lawmakers changed the definition.

Watching and listening to children, we might notice when they work to keep track of their ideas from question to question and day to day. When it seems appropriate, we could look for ways to help them do that.

Technical Vocabulary

That brings us to definitions, which are as important in science as they are in the legal system, and for the same reasons. When trying to reconcile inconsistencies, set foothold ideas, and keep track of those ideas, it quickly starts to matter if meanings are ambiguous. If we're not sure just what the foothold says, it's harder to figure out what it implies, and then it's not very useful as a foothold. We need to be precise.

The need to be precise comes up from time to time in everyday language: Giving someone directions, you might say "take the highway," and in many situations there's really only one highway you could mean. But if there's more than one, you need to be more precise: take the *Kancamagus* Highway. Or, making a New Year's resolution, you could be soft on yourself with something vague like "I resolve to exercise more," or you could make it more precise, and resolve to "run at least ten miles every week." The first one's a much easier resolution to keep, because it leaves lots of wiggle room for interpretation. The second one is clearer: if it's Friday night and you've run only seven miles that week, you've got three to do on Saturday.

For an example from science, think of how we use the words *move* and *moving*. A child fidgeting in a chair is moving. The second hand on a clock is moving. A car driving down the street is moving. But there are shades of meaning that are different from one to the other. The first is a kind of motion that doesn't go anywhere. It would be strange to ask, "Which direction was she fidgeting?" the way we can ask, "Which direction does the second hand move on a clock?" But the second hand doesn't end up going anywhere, either, and asking which way it's moving is different from asking, "Which direction is the car moving?" The car might be moving north or east; the hands of a clock move, well, clockwise!

So to make scientific progress in understanding motion, we need to find ways to be more precise. We have to name the kind of motion. We define terms—they are a kind of foothold idea—and use them to help keep track of meaning. And this is why vocabulary becomes technical—as with giving directions using specific highway names, it's necessary in science to have particular terms with particular meanings. Physicists call the second hand's motion *rotational* and the car's motion *translational*. Technical terminology, in other words, arises out of the need for precision.

Watching and listening to children, we might notice when they come upon that need. We might even notice instances of their trying to address it, such as by talking about or choosing what words should mean. And, when it seems appropriate, we might design experiences in which they're likely to hit upon that need.

Mathematics

If you want to be precise about an idea, you can't do better than to express it mathematically.* Quantify that New Year's resolution, and it becomes quite clear what it means: run ten miles every week. And from this definition you can derive implications: If you've run seven miles, you've got three to go. (If you've run n miles, you've got $10 - n$ left to go!) That's what mathematics is all about—deriving and keeping track of the connections and relationships among ideas. It's just made for the task of developing a logically coherent system of knowledge.

Of course, the mathematics has to make sense. There's no point in science to use mathematics if it doesn't connect to our experience of the world. Probably Galileo's greatest contribution to science was to develop mathematical definitions of motion that connected to common sense. Now physicists define *translational* velocity as the distance moved divided by the time that's gone by; they can quantify how fast. They define *rotational* velocity as the change in angle divided by the time, so they can quantify that meaning of how fast, too. And they can use these definitions to help them understand and define more difficult ideas, including *acceleration*: the *rate of change of velocity*, that is, the change in velocity divided by the amount of time that's gone by.

* Kelvin put it this way, in 1889: "I often say that when you can measure what you are speaking about, and express it in numbers, you know something about it; but when you cannot measure it, when you cannot express it in numbers, your knowledge is of a meager and unsatisfactory kind: it may be the beginning of knowledge, but you have scarcely, in your thoughts, advanced to the stage of science, whatever the matter may be." Sir William Thomson, Lord Kelvin. 1889. *Electrical Units of Measurement. Popular Lectures and Addresses*, Vol 1. London: Macmillan.

That these definitions are mathematical makes them strong footholds, because physicists can use mathematics to derive inferences and build a system of thinking. Mathematics is a powerful tool for science, so powerful that it's easy to misuse. There's so much advantage to getting an idea into mathematical form that there's a big temptation for scientists and educators to get there too quickly. Galileo was always careful to show and maintain the connections between his mathematical derivations and common sense. He showed great wisdom when he held off defining *force* for that reason: he couldn't find a way to reconcile the contradictions in his nonmathematical common sense.

It's a problem that not everyone is as patient as Galileo. Once something is in mathematical form, it's so easy to use it as a foothold—for example, to plug into it as a formula—that it's tempting to hurry the process. The temptation is to get a definition in place and then set off deriving things. In that way, it's easy to lose the connection to common sense. And people tend to give ideas expressed mathematically a higher priority than the ideas might really deserve. Sometimes it's out of intimidation—mathematics can be like fancy terminology. And sometimes it's out of wanting to avoid unresolved ambiguities: plugging into a formula is easier than really trying to figure out what makes sense.

As students come to use mathematics to help them think about the natural world, we can pay attention to whether and how they are using it to express and refine common sense. We don't want it to lead them away from that expression and refinement, because that would be leading them away from science.

Controlling Variables

We've talked about the beginnings of science in the ways people sometimes try to reconcile inconsistencies that come to their attention and how students learning science need to make that a habit. To develop as scientists, they'll need to go further than that. They'll need to make it a habit to *anticipate* inconsistencies.

This involves more shopping for ideas, this time for ideas that someone might one day raise as objections. "Here's my idea," thinks a scientist. "What reasons could there be to disagree with it?" She doesn't just wait for someone else to disagree; she tries to do it herself. She shops for ideas in her own knowledge and experience that might conflict. Like a good lawyer, she anticipates carefully what another side might argue. When she finds something, she tries to figure out how to respond. In other words, the pursuit of coherence in science is all about argumentation.

Here's an example. Suppose scientists want to find out whether living near high-voltage power lines causes cancer. Someone collects data in some region and finds that the people who live near the power lines are more likely to have cancer than those who don't. A novice might think that that answers the question, but scientists know that there are other arguments they'll need to be able to answer. So they need to think about those other arguments. What other explanations might account for the pattern in their data? There are all kinds of possibilities.

One argument is that it might happen by chance, the way if you flip a coin twice it sometimes comes up tails both times. That isn't because anything in particular is making it come up twice; it's just that sometimes that happens. In fact, if you try flipping a coin twice many times, you'll find that about one-fourth of the time you'll get two tails. On the other hand, if you flip a coin twenty times in a row and get all tails, you can be pretty sure it's a trick coin. (The odds of that happening by chance are very small, 1 in 2^{20}.) This is a job for mathematics. The scientists have to figure out *how much* more likely it is that the people near the power lines have cancer, and they have to figure out what are the odds that could happen by chance.

Another argument is that they had some bias in their data collection. How did they identify cancer cases? Did they use the same approach in all of the regions? Maybe the people living near the power lines tended to be insured by a particular HMO, and that HMO was more helpful than others with providing information. Or if the scientists collected data with a door-to-door survey, maybe the person who canvassed the area near the power lines was better able to get people to talk about something so personal as their health than those who interviewed people in other areas.

Other arguments would blame the increased cancer rate on some other cause than the power lines. Maybe it's because the people living near power lines also tend to be living near cancer-causing chemicals. Or maybe people who are willing to live near power lines are more likely to smoke cigarettes or less likely to have healthy diets. As in all shopping for ideas, it takes imagination and perseverance to think up possibilities—what else might be causing this? The scientists need to anticipate those arguments and take those other variables into account. Their analysis has to *control* for those variables—so, for example, they should find out whether the pattern is still there if they look only at people who don't smoke, or people who eat well, and so on.

Similarly, when they conduct experiments, scientists need to try things that will test possibilities someone might raise. So if they think that, say, having power lines nearby affects how plants grow, they could design an experiment to grow plants, some near power lines and some not, to see if it makes a difference. They'd better test enough plants so that if they do find something, they'll know and be able to argue that it's not just chance. They'll need to be sure to treat all of the plants the same ways, aside from putting some near power lines. For example, they need to give the plants the same amount of water; otherwise, how would they know it wasn't the water that caused the difference?

It's all about argumentation. And mechanism, don't forget! The scientists need to prepare themselves to respond only to possible arguments. If they're studying power lines and cancer, they *don't* need to control for the names of people who live near power lines, or their astrological signs, or their tastes in fine art, because no scientist would argue that any of those things could affect whether someone gets cancer. If they're designing an experiment with plants, they don't need to control for what kind of clothing they wear to work, or the language they speak, or what music they listen to while working. They need to control only for variables someone might argue would matter. Those are the variables for which someone might think there's a mechanism: it's plausible that chemicals or smoking or diet could play causal roles in cancer, so the scientists need to check for those possibilities.

When watching and listening to children, it's important to notice when and how they anticipate arguments and try to reconcile them. As with other aspects of scientific inquiry, we might expect children have the beginnings of these practices, too. We want them to learn to control variables, for example; we wouldn't want to miss it if they were starting to do that in their own. (See if you can catch where it happens in the case studies.)

Expert Understanding

We've talked about the very rich common sense of mechanisms in everyday thinking. That's where scientific knowledge starts, and we've talked about this as well: Scientists draw on and refine those resources, moving toward coherent systems of understanding. Trying footholds, they establish principles that help them refine their disorganized everyday thinking into coherent systems of understanding.

Think about the expertise of a chef. Everyone can taste thyme in a recipe, but someone who's lingered and reflected on that taste in many recipes can say, "There's thyme." Chefs can distinguish it from other flavors that are similar but not identical, for example, lemon. Almost anyone could do that in a side-by-side comparison, so this is again something that starts with everyday abilities, but the chefs have refined those abilities. Similarly, chefs come to recognize flavor combinations that come up often in food; for example, they can notice or imagine thyme-and-rosemary all at once and recognize it as a theme in Provençal recipes.

So it starts with an everyday sense of flavors, but the chefs have shopped in those sections of their knowledge and experience often, and they know where things are. Of course, they've also tasted more flavors than the rest of us, so they've add to their sense in that way too, but it would be a mistake to think that's most of what they've accomplished. (Even then, when they taste something new, they connect it to other flavors they already know—tasting anise for the first time, they compare and contrast it with licorice and fennel.)

The same thing happens in science. As students spend time lingering and reflecting on their sense of various mechanisms, as they gain experience shopping in that part of their common sense, they get to be more familiar with what's there. They refine it in the same ways. They can recognize and imagine particular mechanisms more easily. They can keep track of differences among mechanisms that seem similar to people who haven't given them as much careful thought. They can recognize and imagine combinations all at once that come up in many contexts. And of course, they've come upon new phenomena and new mechanisms along the way, so they've added to their sense in that way too, but as with the chefs, it would be a mistake to think of this as most of what they've accomplished. (Even then, too, they first understand the new phenomenon by connecting it to others—on first discovering a certain new phenomenon in light, scientists compared and contrasted it with what they knew about how waves move on ponds and sound moves through air.)

Here's another example of refinement. Think of a large bucket of water with a tiny hole in the bottom. Water drips out, a tiny bit of water with each drip. You know that the surface of the water in the bucket is going down, although you also know you might not be able to *see* it moving down. You know that because each of those tiny drops is taking away a tiny bit of water, a negligible amount on its own, but over time many drops will add up to something noticeable.

And now think about a child who's growing. Measure his height one day to the next and you won't notice a difference, but you know that over time he'll get taller. This is all just everyday thinking, except that now you're comparing two parts of common sense you might not have compared before. You can see what they share: tiny, unnoticeable changes adding up to something significant over time.

That idea comes up again and again in science and mathematics, and so scientists and mathematicians come to recognize, talk, and think about it easily. They refine that rough idea to greater precision, to the point that they can quantify what they're doing. They distinguish different types of the idea: Some kinds of tiny changes stay constant, such as the water droplets from a leaky faucet, coming out at a steady rate. Other kinds of tiny changes don't stay constant: the *changes* change. So the water drips more slowly from the bucket as it runs out of water, or the child gains height more slowly as he finishes adolescence. Like two similar flavors that a chef distinguishes, these are two types of change that a physicist distinguishes: a constant rate of change and a decreasing (or increasing) rate of change.

It goes further. For the faucet, the reason the water drips at a constant rate is that the pressure pushing the drops out stays constant. For the bucket, the reason the water is dripping from the bottom is that it's being pushed through the hole by the water above it. The more water in the bucket, the higher the pressure pushing the water out. As the water drains out, the pressure decreases, so the water drips out more slowly. In other words, the more water in the bucket, the faster it drips out. That's a pattern that comes up often in physics: the amount of stuff determines how quickly the stuff changes. It comes up with rabbits reproducing (the more rabbits there are, the more quickly new rabbits appear); with cooling an oven (the greater the temperature of the oven, the more quickly it cools); and with radioactive material (the more you have, the faster it decays).

Physicists can easily recognize and imagine that pattern all at once, the way the chef notices flavor combinations. They can describe the pattern precisely with mathematics, which they can then use to derive implications and progress to more elaborate patterns of change. The basic patterns become resources for them, the way basic recipes become resources for chefs, which they can then use to understand more complicated things.

In this way, what starts as common sense about how small changes (water dripping, a child growing) can build up into big changes refines into the mathematics of calculus. That's all in the long term, what happens over years of experience, through formal and informal education. It doesn't happen all at once, and no one should expect it to. But it can't happen in the long term if we don't support its beginnings. Our purpose with these cases is learning to recognize those beginnings.

On to the Cases

It's often surprising to adults how easy it is to get children engaged in thinking and talking about questions in science. But it's very easy, at least while they're young. They'll do it by themselves. Of course they will: They're surrounded by the natural world every moment of their lives. How could they not think about it, explore it, talk about it?

It doesn't take school. Parents who are alert to such things will hear their kids and their friends chatting about anything from day and night, to air and breathing, to lightning and electricity, to ice and water. Or they see kids trying the little experiments—not the miming of experiments when they get random liquids and objects and mix them up to make a mess, but simply trying things to see what happens, which the children might not even associate with science. A ball bounces in a funny way off the wall, and they try to figure out how to make it happen again; a rainbow appears on the wall, and they try to figure out where it's coming from; they notice rubber bands making an interesting sound and they see what other sounds they can make.

So what can school do? One thing it can do, unfortunately, is convince children to quit that stuff, at least when it comes to science class, and start "doing it right." That's a well-documented problem for older students: By the time students reach high school, most have learned to separate what they learn in science from their common sense. There are many reasons for this, not just school, but there's good reason to believe that the usual practices of science teaching have been part of the problem. But it's also clear that school can help. Teachers who genuinely attend and respond to children's inquiry can help them learn to use what they know, both about the physical world and about knowledge and reasoning, and build from it.

The next chapter is quick, just a brief introduction and overview of the case studies with some general information and suggestions for how to use them. Then we turn to the case studies themselves, in Chapters 4–9 and on the DVD. You're going to be watching children's inquiries. This chapter has been a primer in what sorts of things there will be to see.

To review: When watching and listening to children, we want to notice (among other things, surely) whether and how they are shopping for ideas and reconciling inconsistencies, all within their common sense of mechanisms. These are the beginnings of science, and as science educators we need to be aware of them in students.

We still haven't talked about teaching, that is, about methods and strategies or even about curricula. That's not because we don't care about teaching! It's because we believe that the most effective teaching is responsive to the students' ideas and reasoning. What we see and hear in children's thinking affects our judgments about methods and strategies, both on the fly during class and in planning. The more perceptive and insightful we are in attending to children's thinking, and the more accurate our diagnoses of what they need, the better off we will be in deciding what to do.

These decisions will come up along the way in discussing the case studies: If you notice something about the students' thinking, what might you then choose to do? That is, they'll come up as specific decisions in specific moments. Some of those specific decisions are discussed in the case studies. But first, in Chapter 3, we offer some general guidelines for focusing on student inquiry.

CHAPTER 3

Using the Case Studies

The next six chapters are the case studies, with each chapter including (1) a quick introduction with suggestions for watching the DVD (except for Chapter 9); (2) the teacher's case study; and (3) notes with specific commentary and suggestions for facilitating conversations about the children's thinking. The full transcripts of all the videos are included on the DVD-ROM. We strongly recommend you print out the transcripts so you can follow along in them as you watch the DVD, taking notes about what you see, and use them for reference during conversations.

This chapter gives a brief overview of the set of cases and some general suggestions for using them.

Overview of the Set

There are five case studies with video and one at the end without. They're presented roughly in order of difficulty, based on our experience in workshops and seminars. Feel free to skip around, but we recommend that you start with at least one or two of the first three, and as you start to get a feel for the game, move on to the fourth and fifth.

"The Pendulum Question" (Chapter 4) is Mary Bell's case study of a discussion in a combination fifth- and sixth-grade class about what would happen to a pendulum she was swinging if she let it go or cut it just when it reached the highest point. The students came up with several ideas and debated them in a lively, good-spirited way.

"Falling Objects" (Chapter 5) is Jamie Mikeska's case study of her first graders discussing which would fall faster, a book or a piece of paper. It includes video from two days, one when Jamie first posed the question for the students to consider and a second day when they did their own experimenting. This is the only case study that includes video from more than one day.

"Chaos in the Corridor" (Chapter 6) is Jessica Phelan's account of a group of eighth-grade students working together to understand the rock cycle, the cycle of transformations rock undergoes from one form to another. The snippet for this case study contains almost exclusively student conversation.

Other than a brief suggestion near the beginning, the teacher does not say anything at all.

The next two cases present examples of student thinking that are more difficult to interpret, based on our experience in seminars. We include them to provide another level of challenge for people who have become comfortable with and interested in watching and listening so carefully to children's reasoning.

"The Trouble with Bubbles" (Chapter 7) is Pat Roy's case study of her third-grade class talking about what shape soap bubbles would come out of differently shaped wands. The students would try blowing bubbles on the next day; in this discussion they talked about their predictions. Much of the challenge is the topic: many students and almost all adults know bubbles will come out round, regardless of the shape of the wand, but it's another matter to explain why.

"The Power of Magnets" (Chapter 8) is Kathy Swire's account of a second-grade class discussion about magnets, specifically the phenomenon of two magnets holding each other through a child's hand. Again, much of what makes the case challenging is the topic: it is difficult to apply common sense of mechanism to the topic of magnetism, both for the students and for seminar participants.

The final case study we include in the collection is "Teaching Chemical and Physical Change" (Chapter 9), Steve Longenecker's account of his eighth-grade students' discussion of whether salt dissolving in water belonged in a category of physical change (like putting glass beads in water) or chemical change (like an Alka-Seltzer tablet fizzing in water). There is no video or transcript on the DVD, but this case study raises an important topic we felt should be represented in this book: How might a teacher support student inquiry while at the same time make progress toward the state-mandated learning objectives?

General Suggestions

We expect that most classroom teachers will at some point encounter some of what we've seen over several years of discussing case studies with educators. Most of what follows boils down to this: *As you watch and read the case studies, please focus first on interpreting and appreciating the substance of what the children are thinking.*

Many people are tempted to focus instead on what the teacher is doing rather than the children, or on what the children *aren't* thinking rather than what they are, or on general features of the children's behavior rather than the substance of what they are saying. These are certainly important topics, and we do spend time on them in our workshops and seminars, but our core purpose in this collection is attending to children's ideas and reasoning. So for the rest of this chapter, we will elaborate on five general suggestions:

1. Temper the impulse to evaluate the teacher.

2. Focus on understanding the students' ideas and reasoning.
3. Support interpretations with specific evidence from the case.
4. Recognize but tolerate incompleteness and uncertainty.
5. Watch the video before reading the teacher's account.

In addition, each case comes with its own specific, detailed facilitators' notes. Please consult them as well, particularly if you are preparing to lead a seminar.

Temper the Impulse to Evaluate the Teacher

The greatest challenge we've found to productive conversations about cases is that people feel compelled to evaluate the teaching.

Usually the evaluation is critical, and often after only the merest of glimpses into the case. In our workshops it happens again and again, unless we do something to forestall it: we show a three- to five-minute video clip, and the participants in the workshop begin finding fault. The teacher has ignored one child or focused too much on another; the teacher hasn't had the students conduct an experiment or clarified the question or given the students enough background information; the teacher should first have them talk in groups, or draw diagrams, or record observations. Some of the criticisms people raise contain valid points to consider, but there are several reasons to be cautious.

Quick Criticism Is Often Naïve

To an observer, it can seem perfectly clear what the teacher should do, but that's almost always because the observer has limited information. If it seems obvious watching the video that the teacher should call on the girl who's raising her hand, it could be because we don't know that girl; maybe she's *always* raising her hand and tends to monopolize discussions, and the teacher is trying to draw out a student who hardly ever participates. Someone's opinion that the teacher should use this technique or that, from a general sense of best practices, may not be justified for this particular class for any of a number of reasons. Or maybe the teacher *has* used that technique, just not in these three minutes of video.

Rather than cast judgment, we should cultivate habits of asking why the teacher's choices might have seemed appropriate to him or her. What might the teacher have been seeing and thinking, and what is the critic seeing and thinking that differs? What perceptions and interpretations of the students' reasoning underlie the teacher's actions and the critic's concerns? It's more difficult work to imagine what things might have looked like from the teacher's perspective, but it's important to do.

Even when we have much more information, particular choices in particular classes are always uncertain. Maybe if the teacher had chosen to conduct the experiment more quickly, the students would have lost out on the rich discussion they were having; maybe if she'd organized them into groups, some would not have known what to do; maybe if she'd given more clarification, it would have shifted their attention more toward her thinking than their own.

So instead of drawing conclusions about what the teacher *should have done*, try to think in terms of possibilities, what the teacher *could have done*, recognizing all along that none of us has any way of knowing what the effects of any choice would have been. We speak of coming up with menus of possibilities for how teachers might respond; the metaphor of a menu helps us remember that there's always more than one reasonable way to proceed.

Quick Criticism Discourages Teachers from Letting Anyone Examine Their Practices

The second reason to temper the impulse to evaluate is that it can have an effect on you and those around you. Because teaching is an "uncertain craft," as Joseph McDonald (1992) put it, teachers who are brave and generous enough to let others examine their practices are setting themselves up for disparagement. It's not an easy thing to do, and sometimes it's the people least willing to do it themselves who are the most judgmental. Speak about the case study as if the teacher were in the room, because *teachers are in the room*, and they shouldn't leave the conversation thinking, "Whew—I'm never letting anyone watch me!"

We recommend this for individuals as well, reflecting on their own teaching: Don't judge your actions by unattainable criteria. It is not possible to know how students will respond, so lessons won't generally go as planned. Nor is it possible to follow every opportunity that comes up in class, so some will be lost. If you torment yourself for everything you missed or every decision that, in retrospect, you would have made differently, then examining your teaching will be painful for reasons that simply aren't valid. Looking back, you will *always* see things you hadn't noticed in class and think of other ways you might have acted. That's good! It will give you things to consider in planning ahead.

Evaluating the Teacher Draws Attention Away from the Students

We've suggested it's important to be careful and respectful about criticizing the teacher. But for our purposes with these case studies, we also think it's important to be careful about *praising* the teacher. We're trying to focus on the substance of the students' thinking. To that end, even favorable evaluations of the teaching are mostly beside the point.

In truth, we think these are all examples of excellent teaching, and we're very proud of them. But as much as we'd like to celebrate what the teachers are doing, we want the focus of the conversations to be on what the children are doing. In principle, our purposes here would have been served as well by cases in which teachers made poor choices, such as to dismiss or ignore students' incorrect ideas, as long as the data included evidence of those ideas. That's why we try to redirect assessments of the teacher, even favorable ones, to the students: What perceptions and interpretations of the students' reasoning underlie the teacher's actions and the observers' assessments?

Even if they're not being judgmental, the first things people in our seminars usually notice are the teacher's strategies. We expect many readers will do the same—start out noticing the teachers and their methods, including science talks as an approach. There's nothing wrong with it if the case studies give you ideas for strategies.

For this book really to be of value, though, you'll need to shift your attention from what the teacher is doing to what the students are doing; and if you don't, you'll be disappointed. There are elements of method in the case studies, and you'll get some ideas, but we didn't design these cases to tell about teaching. The teachers didn't collect snippets that would demonstrate methods; they collected snippets that show student thinking. So, for example, you'll see more full-class discussions in the cases than you would if you just dropped in on the classes, when you'd be just as likely to see another sort of activity (students writing, experimenting in small groups, discussing in pairs, and so on), for the simple reason that class discussions are a very efficient way to hear children's ideas in the space of a fifteen- to thirty-minute video clip. It's not because we think all science teaching should be in the form of science talks!

Of course you're going to want help with thinking about those other activities; there are many excellent books and websites that could help you with that. We're trying to do something else for which, we believe, there isn't already an abundance of materials.

So try to focus on the students. It might help to think of your ideas about what the teacher should do as arising from interpretations about the students. For a simple example, when you think "she should have taught them about X," it's probably because you've seen something that makes you think they don't know about X. So talk about that—what you've seen and how you've interpreted it. Then you can move on to talk about ways to respond.

Focus on Understanding the Children's Thinking as Inquiry

Try to focus on the sorts of things we discussed in Chapter 2: the children's sense of mechanism, their shopping for ideas, and their attention to consistency. How are they using and looking for evidence? How are they reasoning from what else they know? What sense are they making, and how are they articulating it?

For many people, paying that kind of attention makes a lot of sense, in the abstract; they recognize that as what science is all about. But when they get down to cases, they focus instead on other things: How are the students behaving? Of the students in the room, how many actually speak or look like they're paying attention? These things are certainly important—and of course you'll see them and you'll talk about them. Our experience, though, is that most people have an easier time seeing these aspects of what's happening than they do thinking about the particular things children say. So when

Ashley says in the first case that the pendulum is "gonna, like, fly the opposite way," (line 66), don't let yourself off the hook of trying to figure out what she might be thinking just because some other students don't seem to be listening. Our purpose in choosing these cases is precisely to help you get practice in understanding what children say. You're absolutely right to care about those other students! Just be sure to use the data you have available, in that moment, to think about Ashley's reasoning.

Another challenge for many people is to focus on what the children are thinking beyond an assessment of whether or not they are correct. Some with more experience in science education get caught up in the game of hunting for misconceptions, as we discussed in Chapter 2, attending to student thinking in order to find its flaws. They notice the children's ideas, and they compare them to scientists', looking for similarities and differences between children's and scientists' conceptions in various areas rather than at how students are engaging in inquiry. Sometimes, too, people want to talk about what the children are *not* saying or thinking that they should be, rather than try to understand what they *are* saying and thinking.*

So we are suggesting you make a concerted effort to understand and appreciate the *students' sense* of the matter. When a student says something that sounds strange, the challenge is to understand what he might mean by that and why he might think it is reasonable. Sometimes the best interpretation will be that he was just joking, or he was only mimicking knowledge he didn't really have, but sometimes it will be that he's onto something wonderful. One way we try to help people get started is by suggesting they imagine the children are *brilliant* and so we should hang on their every word, trying to understand their meaning. So when Ashley says the pendulum will fly the opposite way, try taking it seriously in the way you would if you knew her to be gifted: What might she mean?

Support Interpretations with Specific Evidence from the Case

In making an interpretation of the students' thinking, make a point of identifying the particular statements or aspects of the students' behavior that support that interpretation.

Often people's ideas about the students come from general expectations rather than from the case study itself. Much of the value of using case studies, however, is in focusing on the evidence at hand. Sometimes it's at odds with our expectations. In fact, it's standard experience for us, when running workshops on cases in this project, that participants' expectations are challenged by what they see.

It's not unlike what we hope children will do in science: check your expectations against the data at hand, and be ready for the possibility that

---

* It does happen, sometimes, that the key to interpreting students' thinking is realizing that they're not making the same assumptions adults take for granted. In those cases, it is very useful to talk about what they're not thinking.

they will not match. During the project, we came up with a prompt sheet to help get things started; we've included that on the DVD-ROM, and you might find it helpful. In our conversations we focus closely on specific things students say, and it's not at all unusual or inappropriate to spend ten minutes or more talking about a single statement, such as Pat Roy's student Zoë's claim that a bubble "can't get all the sides that it suppose to get" (line 69 in the transcript for Chapter 7 "The Trouble with Bubbles").

It may seem strange to devote so much attention to a student's wording. When teaching a lesson, you clearly can't stop class to go analyze every utterance. Nor, certainly, can you rewind and listen to it again, to see if maybe you misheard. But that's the whole point—to do something you can't do during class. A team of medical school students will do similar things to analyze a piece of information from a medical case: they'll debate its meaning and significance to the diagnosis, in ways they won't be able to in practice with real, present patients, whom they'll see without any colleagues and then only for the few minutes of the appointment.

We also like to compare what we do with student thinking to what scientists do with specimens. No one would find it strange to see a geologist poring at length over a rock sample that's abundant in some region, an entomologist over a moth, or a botanist over a leaf. Examining a specimen closely can help them develop new understandings. That's just what we hope to achieve in examining moments of student inquiry.

Typically, we find that people don't notice very much when they are first getting started. As they progress, they notice more and more and gain new appreciation for children's ideas, reasoning, and abilities.

Recognize but Tolerate Incompleteness and Uncertainty

The first part of this suggestion is to remember that the case studies tell you only so much. The video portions vary from fifteen to thirty minutes, which is a small sampling, which is why we call them snippets. The teachers provide background information about the classes, but there's only so much they can include, even about the preceding hour.

We should also note that in some instances we chose to omit information some might consider relevant. In general, the case studies say little about what happens *next*—people often want to know how things turned out in the end, by which they usually mean "Did the students get the right answer?" In our workshops, having that information—which of course was not available to the teacher at the moment in the snippet!—tends to pull the conversation back toward evaluating the teaching, and evaluating it by the traditional criterion: if the students got the right answer, the teaching worked. So we often just leave that information out.

Another kind of information we omitted for ethical reasons. When we were preparing these materials, it was important to all of us that students in these classes not suffer for having appeared in them; we'd like to think that some of them might one day read these cases and watch themselves on video, and we edited with that in mind. So, for example, we left out information such as that a particular student had a learning disability, or was diagnosed with ADHD, and so on. That is a small compromise on the value for the reader, because it puts you at a disadvantage in interpreting what's happening, but it was necessary. When (as we hope) you try collecting snippets from your own teaching to show colleagues in private, you'll be able to include those details.

In all, the first point here is that it's important to recognize that there is only so much information available in the case study. More might help you interpret a student's thinking or understand the teacher's interpretation. And, again, even if you had much more information, that wouldn't eliminate the uncertainties of interpretation.

The second part of this suggestion is to proceed regardless. Early in the process, some people balk at the idea of interpreting what a student meant, because they feel it's hopeless: "How can we possibly be sure?!" Part of the answer to their concern is that the same is true in science, where we're used to the idea that we can't expect certainty. It's OK not to be sure! Another part of the answer is that it's often surprising how far you can get in making sense of a student's thinking from the evidence that's available. And still another is that there's much to be gained from lingering and mulling over a student's comment just to consider the possibilities for what it might mean, in particular to discover possibilities beyond what you first noticed. Maybe it wasn't a ridiculous thing for the child to say; maybe there was something to it. And if you're talking about the case study with colleagues, the main goal is to understand others' interpretations; it's not important that you all agree.

 ### Watch the Video Before Reading the Teacher's Account
We end with the easiest suggestion to implement. Read the introduction to the case study to get background information about the class, but hold off reading the rest of the case until you've watched the video yourself.* That will let you form your own impressions and come to your own interpretations, to practice seeing and hearing children's inquiry, which of course is the main idea. If you can, watch the video with colleagues, so that you can hear a range of impressions. That's what we do in our workshops and seminars: we show the video or a segment of it and ask participants to talk about what they saw in the children's inquiry.

Then go back to the case study and read what the teacher had to say about it, what she was thinking during class, and what different ideas came up later. You'll almost always see some things differently and some the same; there's no one right set of things to have seen. As we've said, it's OK to disagree, but try to understand the rationale behind the other interpretations.

As we think you'll see, or you may already know, video recording adds greatly to the sorts of discussions people can have about classroom

* Except, of course, for Chapter 9.

interactions. The camera doesn't catch everything, not by any means, but it does provide a rich and immediate sense of the class. You'll be able to hear the tone in students' voices and see the expressions on some of their faces. It's not the same as being there, but in some ways it's better: you can play and replay the video, follow along in the transcript, and stop the action whenever you like.

Watching the video on a DVD player, you'll have the option of using the subtitles or not, just as with foreign language movies. Or you could read along in the transcript.* The DVD files are divided into scenes, as you'll find on any movie you rent, and we've noted the divisions in the transcripts.

You'll notice in some places that a student's face is blurred, and in one case ("The Trouble with Bubbles") a student's voice is dubbed. These are students we did not have permission to include in published video. We're often asked how we chose cases for the collection; one of the principal criteria was that there not be too many students present for whom we did not have permission, because it is a fair amount of work to blur their images, and because too much blurring is distracting.† (During the project, as a matter of principle, the teachers did not treat students any differently if they gave permission or did not, so much depended on luck. Fortunately, we usually had only a couple of students without permissions in any class.)

You'll also notice that we're not professional filmmakers! In most cases, a member of the project staff visited the class and taped it with two home-quality digital cameras. We then imported and edited those tapes on Mac computers, with iMovie or Adobe Premiere. Even if we were professional photographers, or could hire them, we might have chosen the minimal approach we've taken, to be less intrusive in class and less "produced" in the final product. If you're moved to try taping yourself, that's great! That's how we got the first day of "Falling Objects": Jamie videotaped the class herself. We'll give a few tips on how to go about it in Chapter 10.

---

* You may notice some differences between the transcript printed from the DVD-ROM and the captions onscreen, especially for "Chaos in the Corridor," where, true to the title, the students were often speaking over each other, and there was too much to caption effectively. And for either, it's likely if you watch closely enough you'll catch some mistakes. We've tried to be as accurate as possible, but transcribing is not as easy as it seems!
† There were many other considerations as well, from whether teachers had time to prepare the written case study, to the sound quality in the video, to, simply, our sense of which case studies have led to the best conversations in seminars.

CHAPTER 4

Fifth and Sixth Graders Discuss a Dropped Pendulum

This was the first case developed in the project—in fact, it was the very first thing we videotaped—so we're especially fond of it.* It was also Mary Bell's first time engaging her students in a science talk. She used a question we had discussed during our summer workshop. We were inspired by how quickly and well the children began having this kind of discussion.

Mary's case study begins on the next page. We recommend that you start by reading only the introduction to get some background information about the class; stop when you see the heading "Pondering the Pendulum." Then watch at least some of the video.

Suggestions for First Viewing

Be sure to print the transcript "The Pendulum Question" from the DVD-ROM; it's useful for following the video snippet (especially in this case, when the voices are sometimes hard to hear), for jotting notes, and for reference in conversations about the snippet with others.

The snippet is about half an hour long. When we use this video in seminars and workshops, we stop it at various spots in order to encourage participants' close, careful examination of student thinking. You might find that helpful, too, especially on your first viewing. We generally pause for reflection and conversation in three places:

1. *After Mary poses the question (lines 1–6 on the transcript).* After pausing here, we invite participants to spend a few minutes talking about what they think about the question itself, that is, what they think will happen to the washer. (Sometimes we skip showing the video of Mary posing the question and simply ask it ourselves. We then start the video with Chris starting the discussion at line 11 on the transcript.)

---

* But please forgive the quality of the sound: we had not yet switched to using directional microphones, so some of the children's voices are hard to hear.

2. *After Jeff says, "I think Victoria wants to say something," about five and a half minutes into the video (line 79).* We usually stop here and ask participants to think about what's happened so far and write notes on and/or discuss the following questions with others: What do you notice about the students' reasoning? What are they thinking? What aspects of their thinking might be the beginnings of science?

3. *After Depo says, "If it's just got to its highest point and you just cut it, it would, it will just go straight down" (lines 149–50), about ten minutes into the video.*

In most workshops, that's as much of the video as we use. That may seem surprising if you're just getting started at this, but as we've said, it always seems surprising to students how much more scientists can see in a specimen than they do. The idea is to learn to see; when in doubt, look a little longer.

When you feel you've seen enough of the class, read Mary's account and compare your thoughts and impressions with hers. Finally, if you're interested to read more or to prepare for leading a workshop, read the facilitators' notes on page 58.

■ ■ ■

The Pendulum Question

Mary Bell, Prince George's Public Schools,
Prince George's County, Maryland

As a special educator for twenty-seven years and a Reading Recovery teacher for four, I am well versed in pondering student thinking. Studying how students think about science, however, was a new challenge. During our summer sessions, my colleagues and I had discussed science through numerous open-ended questions. Would students think about science in the same ways we had? Would they use their experiences to support their ideas? Would they change their minds after hearing other students' ideas? Would their ideas be based on plausible thinking?

I teach in a large suburban elementary school in Prince George's County located just outside the nation's capital. In my current position as a resource room teacher, I often coteach with colleagues in my school, visiting their classes for special purposes. This case study focuses on a class I cotaught in November 2000, a combination of seventeen fifth graders and eleven sixth graders. Twelve had been identified as "gifted," although I was not yet aware of who they were, having recently been reassigned to work with this class. I was already responsible for teaching the sixth-grade students in my resource room, but four days a week for one hour the students were all together for science.

Posing an open-ended question and allowing conversation to flow was a new and interesting concept for me. Correctly answering the question was not an immediate necessity, although of course I hoped eventually the students would get there. My first goals were that they present their ideas clearly, hear and respond to each other, and support their thinking with life experiences. By studying what they had to say, I hoped to learn more about what ideas they had and could support and whether they could understand other points of view.

Participating in this kind of conversation about science was new for the students, too, so I felt it was important to spend time talking about the goals. I used the first science talk to set the stage, telling them that we would be discussing scientific questions and that the answers to the questions might not be brought up and in fact might not be known. The important thing was that we support our thoughts with examples, respond to each other, whether to agree or disagree, and, in general, have fun with science. I suggested some questions we might discuss, such as Why does ice float? and told the students they would have the opportunity to generate open-ended questions. Several students suggested Why did the dinosaurs disappear? I then had the class make up governing rules, a technique suggested by one of my colleagues in the project, which helped the students comprehend the format of our science talks. They came up with a nice set of rules mostly about having good manners, such as *Be respectful of others* and *Don't interrupt others while they are speaking*.

The fifth-grade students had recently been studying pendulums as part of the regular science curriculum, composed of performance-based instruction in the form of tasks similar to the activities of the Maryland state assessments given in third, fifth, and eighth grades. This task was to investigate a pendulum made from a string and a washer to determine how the length of the string affected the swinging rate: the shorter the string, the faster the rate of the pendulum swing; the longer, the slower. The lessons involved making predictions, some experimentation, and, very important for the assessments, writing reports with conclusions. Although the students had worked in groups at times, they had not had much conversation. The sixth-grade students had followed the same curriculum the previous year. They were studying air pressure, and their task was to figure out, through experimentation, the best proportion of baking soda and vinegar to fuel a rocket.

During our summer workshop, my colleagues and I had discussed a different question involving a pendulum, and I was curious to see how the children would do with it. We teachers had taken almost three weeks and hours of discussion to agree on an answer. Would the children reach the same possibilities? Would they in any way parallel our discussion?

As I entered the room, the children were visibly excited. This would be our first official science talk—the previous one had been for practice—and there was excitement in the air! (This excitement never left as the year progressed.) I was a bit nervous and at the same time filled with anticipation. What would happen? Would the students talk? What would they think? Would they follow their rules?

Pondering the Pendulum

I had a washer attached to kite string, and I gave the students journal writing paper. The top half was blank for illustrations, and the bottom half was lined for writing. I posed the question as I rocked the washer back and forth on the string.

> If I swing this pendulum back and forth and then cut the string when the pendulum is at its highest point, just before it gets ready to swing back, what will happen to the washer?

I asked the students to think and then write or draw their predictions on the paper. After a few minutes, I had them talk about their ideas, using their notes but not reading from them.

Throughout the conversation, I tried to keep the following thoughts in mind: Are the students presenting clear ideas? Are they providing evidence for their conjectures? Are they recognizing different perspectives and comparing evidence for and against these perspectives? Thinking on my feet, I had to make quick decisions as to when to remain quiet, when to probe, and when to simply restate an idea. I wanted to facilitate their thinking, but I also wanted to provide scaffolding toward better scientific inquiry, and possibly toward the correct answer.

Because I was just starting with some of these students, I had them say their names before speaking. For ease of reading, I have left that out of the transcript here. I include, in parentheses, the grade level of the students.

I let them keep their papers and asked, "Who would like to begin?" Chris was the first to speak.

> **Chris (6):** I think that when you, um, when you let it go, and it's facing a direction, it'll fly, um, it'll fly that way and then it'll go down. Like if, if it was facing the, um, right, and you let it go, it'd be going right for a while.

I found it difficult to understand exactly what Chris was saying, but his explanation sounded like a possibility we had discussed during the summer. I decided to have him draw his idea on the board, starting him out with a diagram of the string and washer that showed the point where the string would be cut. Chris drew a line at an angle coming down from the washer and toward the direction the pendulum had been swinging, roughly as the following illustration shows. Immediately I was pleased with my decision to use the board. In front of me was one of the possibilities we had discussed during the summer.

> **Mathew (5):** I agree with Chris because I think that, that when the force comes, when the force comes, or if you cut it or let go of it, it will go flying like, like this.

Chris' Idea.

Mathew had used the term *force*, and I wanted to know what he meant.

Teacher: So what's the force that you're talking about?

Mathew (5): Gravity I guess; I don't know.

Teacher: OK, gravity is the force that's gonna make it fly out and come down?

Mathew (5): No, I was just guessing.

Ashley (5): Gravity is pulling it down.

Mathew did not want to commit to his idea, but I thought he was thinking of the movement of the pendulum itself, like Chris, not really of gravity. Ashley came in and clarified that gravity would pull the washer down. I hoped someone would continue that line of thought, but for the moment no one did.

Shadawn (5): I think like, um, like, the same thing that Chris and Mathew said. Like, when you cut it, it will fly, it will fly somewhere, whatever. It, um, I think it's like the air, that when it goes, I think, like, the air . . . is pushing it.

Rebecca (6): When you're swinging it, the force of swinging it, and if you let go, then it'll fly the direction that you were swinging it and it'll fall to the ground. [*Makes a hand motion similar to Chris' drawing*]

Ike (5): I think if you're swinging it and you let it go, . . . it'll probably go in the air and go down. [*Ike's eyes glance upward, leading me to believe he is thinking of another possible path.*]

Teacher: OK, so you think . . . something different's gonna happen?

Ike (5): Yeah.

Teacher: Come on up and we'll let you use a different color chalk; why don't you use green. This is the washer. You show us where you think it's gonna go.

Ike's Idea.

I had not planned on using the chalkboard, let alone different colors of chalk, but I felt that doing so would clearly delineate the choices. Ike drew a line that curved upward from the washer and then arched toward the ground. Once again, I was seeing a possibility my colleagues and I had pursued during the summer. I was extremely pleased with how the conversation was going and was quite curious as to what the students would say next.

> **Ike (5):** I think something like that.
>
> **Teacher:** OK. And why is that?
>
> **Ike (5):** Because once you're swinging it, and you let it go, it'll go in the air because . . . it probably depends on how you swing, which way you swing it.
>
> **Mathew (5):** I disagree with Ike because, well it really doesn't matter. . . . If you let it go, if you let it go to, like, down here to this side, it will go flying there, but if you let, if you, um, cut it at the very top of it, it'll go up and then come down, at the very top.

Mathew's gestures made me think he was confused about where the pendulum would be cut or let go, so I restated the question.

> **Teacher:** I'm cutting it right at the top, right before it's getting ready to come back the other way.
>
> **Mathew (5):** Never mind, never mind. I disagree with that.
>
> **Teacher:** What made you change your mind?
>
> **Mathew (5):** It depends on how fast it is going. . . . And how short the string is.
>
> **Teacher:** So, you think it is gonna depend on how short the string is and how—
>
> **Mathew (5):** Not too short though.
>
> **Teacher:** —and how fast the pendulum is going?

Mathew had clearly been listening, and now he seemed to be tying in his experience with the fifth-grade science task: how long the string was and how fast the pendulum was swinging would affect what the washer would do. He was still thinking, though, not quite sure of his answer. Ashley and Mathew then had a short dialogue about the path of the washer, which I do not include here, but it again led me to restate the question. When I did, Victoria came up with a whole new idea.

> **Teacher:** OK, if it's right at the top and it's about ready to come back down, which way do you think it's going to go? [*Various students gesture.*] Do you agree with this then? [*Teacher points to the board.*]

> **Victoria (6):** I disagree with all of them because I think that if you cut it at the top, then it is, like, gonna kind of curve and then come straight back down.

> **Teacher:** You wanna draw what you think, Victoria?

I called on Amber while Victoria went to the board.

> **Teacher:** Amber, do you agree with her?

> **Amber (5):** I agree with her because . . . if it's going fast or something, and, and then you cut it, it might curve around because it was going so fast and the speed allows it to, um, and the speed allows it to come back up and go up so forcefully.

Victoria drew a straight line from the washer to the ground.

I was amazed. Here was the correct answer and we had only been discussing the question for less than ten minutes! I had to keep a poker face and not show my excitement. Would the other students embrace this idea or try to refute it?

> **Teacher:** OK, Victoria, and correct me if I'm wrong, Victoria—you're showing that you think that once we cut it, it's going to come straight down.

> **Victoria (6):** I think that, that it's gonna, like the string, like, gravity is gonna push it down and the string is gonna kind of curve and then just come straight down to the bottom.

> **Teacher:** OK, so the string might curve a little bit but the washer is going to come straight down.

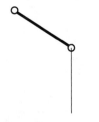

Victoria's Idea.

I questioned Victoria mainly to make sure the class understood what she was saying. She had said something about curving, but her drawing went only straight down. And Amber's statement, although she said she agreed with Victoria, was confusing to me and possibly confusing to the other students. Victoria responded nicely, saying that gravity would make the washer go straight down. She also explained that the string may do something different and be affected by the washer.

Other students had been busy shopping for ideas, and there was a lot of talking at once.

> **Jeff (5):** I agree with Victoria because the washer is a lot heavier than the string, and the washer will come straight down.

> **Brandon (6):** I disagree with Victoria . . . and I kind of agree with Ike because if the washer, if you swing it, and then at the top point, it's gonna go flying up some and then it's going to drop down.

> **Shadawn (5):** I kind of agree with, um, Amber because, like, it depends on how fast . . . it's going. Because I think, like, . . . if it's going really fast and you cut it, it's gonna fly somewhere and do all the curves and stuff, but if . . . it's, like, going really slow and you cut it, I think it's just gonna go straight down.

Jeff had been listening, and he restated Victoria's thinking. Brandon and Shadawn, on the other hand, went back to earlier ideas. Brandon's statement, "I disagree with Victoria and I kind of agree with Ike," showed he too had been following the conversation, although this was his first comment. Shadawn took her disagreement a step further, trying to reconcile the different ideas by suggesting that the speed of the pendulum would affect where the washer would land. I was very pleased with the students' thinking.

But they were not supporting their thoughts with personal experiences. Responding to Shadawn, I tried to encourage them to move in that direction.

> **Teacher:** Shadawn, can you tell us why you think that? Is there something that you know of that maybe makes you think that?

Shadawn only shrugged, but Mathew jumped in.

> **Mathew (5):** Can I say something? I agree with, um, Shadawn because it's kind of like you, you have . . . , like you know how sometimes on movies and things and real life, they have lakes or swimming pools and you have a little rope and you run and grab onto the rope and then fly and then let go and you go flying over to the side? That's just like that, the washer.

I wanted to yell, "Yes! Yes! That's what I mean." That was just the sort of personal experience I wanted in the discussion. As the room filled with more conversation, I kept my cool and focused on Vanice's concern about Mathew's idea.

Vanice (5): Not exactly, Mathew, because the pendulum is, I mean the washer is tied to the string so it won't go to the other side.

Mathew (5): But she is cutting it, or she'll let it go.

Vanice (5): But it is still not going to go to the other side because it's hooked together. If it wasn't hooked together then, yeah, it might go to the other side. . . . The string would still be in your hand but the . . . washer will go somewhere else.

Mathew (5): I know that. It's kind of like, it's kind of like, um, the person flying off of it letting go and then going into the water.

Vanice (5): I know, but it's not connected, I mean, it's connected, so that wouldn't work.

Once again, I was taken aback. My summer colleagues and I had also discussed this very same example of jumping off a rope swing, but we had not been so quick as Vanice to see this difference, that the washer would not separate from the string as the jumper would from the rope.

Then Grace came in with more ideas.

Grace (6): Well, I agree with Chris because, um, it can't really go up more because, like, gravity doesn't go up. . . . I don't think it can just go straight down because, I think, you're swinging it.

Mathew (5): I disagree with Grace because, because it's kind of like you throw a bucket or a ball up in the air, gravity is coming down, forcing it to come down, but . . . it [is] still going up.

There was a lot of talking as Depo took the floor to argue with Mathew.

Depo (6): I disagree with Mathew because, like, gravity always pulls you down. If you throw it up, . . . that's your force, but if you just leave it alone, it will just come down.

Mathew (5): That's what I'm trying to say, that it's the person who pushes it, how fast it goes, that it determines how high it goes.

Depo (6): If it's just got to its highest point and you just cut it, it would . . . just go straight down.

I was very impressed. The students were listening to each other, using experiences to support their ideas, and making sense! Mathew brought in another example from experience, and this time he used an experience separate from the pendulum idea and swinging—throwing a bucket or a ball up in the air. It is interesting to note the class' reaction when an idea or experience with common ground was expressed—there was a burst of new talking. That was a good sign that the students were engaged and thinking.

And they stayed engaged and thinking for the rest of the period. Victoria brought in her experience with swinging keys, and other students took up that example—Shaquai, Brandon, Jeff, and Ashley. As the conversation went on, students were paying more attention to how the path of the washer would depend "on the speed," "on where you let it go," "on how high it's going," and "on the weight and the gravity." It seemed especially difficult for them to think of letting the washer go at the highest point of its swing, so I frequently restated the question, trying to help them think about that moment when it was "just about ready to come down again."

Other students brought in more examples from experience. Grace talked about jumping off a playground swing, another example my colleagues and I had discussed in the summer. Shadawn talked about her experience walking along, "swinging her stuff and it's really heavy" and how one time she "let it go and it fell on the curb." Brandon compared that to a crane with a heavy load.

The room was full of children talking, remembering life experiences, supporting their positions, and listening carefully to the conversation. I was pleased to see how tangible their reasoning had become, how they were swinging imaginary pendulums in the air and trying to picture what would happen, swinging real pencils and pens, and trying to reconcile differences and reach a plausible conclusion. Toward the end of the discussion, they were coming to the conclusion that it would depend on the weight of the pendulum, the height of the swing of the pendulum, and the speed of the swing. (Mathew said that he had changed his mind as a result of the talk.)

With ten minutes left to the school day, it was difficult to have the students experiment with washers and kite strings. I asked them to write what they thought would happen on the back of their original prediction. As I collected the papers, the students begged me to try the experiment or tell them the answer. During the summer workshop, my colleagues and I realized how difficult this was to do. (It really does depend on where you let it go! And if you don't let it go just at the end of the swing, at the instant when it isn't moving, then you won't see it fall straight down.)

Reluctantly, but with the students' eagerness, I made one attempt using the washer and kite string, and I helped the students see that the washer would fall straight to the ground. Unfortunately, we did not have the chance to go back and talk about what this meant for all the other explanations.

Reflections

The students in this science talk were engaged and thinking. Comparing it with my previous experience, I was delighted with the results. I would have liked to see more students participate in the discussion, however. Perhaps as the facilitator, I could have encouraged children to speak by giving positive feedback and praise on their written predictions. For example, I could have encouraged them with a comment: "This is a wonderful idea. You might want to share it with the group next time." I could have also provided a chance to use the think-pair-share strategy, where students meet with a partner, talk about the question, and then share their ideas with the group.

In the future, I would like to see more evidence of students' thinking about opposing positions. I could extend the talk by dividing the class into thirds and asking each group to find ways to defend one of the pendulum paths. This would encourage them to look for inconsistencies and try to reconcile them, whether or not they believed in that idea. More students would become involved and perhaps other experiences would filter into their conversations.

After examining the students' written predictions, it is interesting to note that before the science talk, five students believed the pendulum would fall straight down, twenty believed the washer would fly up, make a curve, and then come down, and one student wrote something inconclusive. After the talk (but before the experiment), seven students wrote that the washer would come straight down and nine that it would depend on the height, speed, and weight of the pendulum. Five students thought the washer would fly at an angle away from the direction of the pendulum, and five wrote that it would fly up and then down at an angle. So the conversation had a big effect on their thinking.

As I said before, I was amazed at the way the students' discussion paralleled our summer session. In forty-five minutes, they had covered our possibilities and even used many of the same examples to try to find a solution. Would this continue? How could I move the students into deeper thinking?

As the year progressed, the students grew by leaps and bounds. Mathew, in particular, blossomed and continued to participate enthusiastically in many of our discussions. He was not one of the students identified as gifted, but his leadership during our science talks overflowed into other subject areas and changed his peers' perception of him. At the end of the year, he won the class award for most improved student.

The students used more and more of their everyday background knowledge to support their thinking. We continued to use the blackboard for diagrams and also began to keep track of ideas by making abbreviated lists. The students learned to try not only to prove ideas but to disprove them as well. Freely using the dictionary, blackboard, and teacher-provided manipulatives, they discussed and pondered many questions. More students began to contribute orally, and the fun never left our science talk sessions. On the contrary, it became difficult to teach under time constraints and curriculum demands, as students wanted to discuss concepts and learn through oral inquiry on a daily basis. They established a mindset of thinking of multiple possibilities in answering a question. It was a nice kind of difficulty to have as their teacher.

■ ■ ■

Facilitators' Notes

Please see the general notes for facilitators in Chapter 3. Here we'll provide specific comments and suggestions with respect to discussing the case at the recommended stopping points, the rest of the snippet, and the teacher's case study.

Our purpose here isn't to present a thorough analysis of the snippet but to give a sense of some possibilities for topics that might arise and topics a facilitator might bring up.

What Might Students Say Will Happen to the Washer?

It often happens that people are reluctant to stick their necks out to say what they think will happen to the washer; sometimes there are physics teachers in the group, and they tend to have ready answers. So we often ask participants instead to talk about what they think students might say in response, which lets the physics teachers contribute in a different way and lets anyone make a suggestion without the risk of being wrong. (Of course, we'd rather that weren't such a terrible risk to take! But very often it is a risk, and it helps to have a way to make things easier.) Often we have to decide whether to guide or provide the answer to the question, as part of or after this conversation. It's a tough call! For some people it is distracting to be unsure of the answer, and they have trouble paying attention to the students. For others, it seems to work the other way: not knowing the answer, they pay closer attention to the substance of the students' arguments. We give the answer later in the chapter, in case you missed it in Mary's case study and would like to see it. It's in the part under the heading "The Physicist's Answer."

Most Conversations Touch on Options Similar to
Those Mary's Students Considered
Someone always has the idea that the pendulum will go out, or up and out; often someone says it will fall straight down. Other answers that may come up: outward along the line of the string, back inward toward the center, as

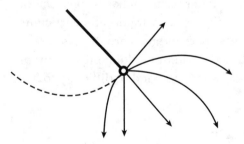

Answers That Might Come Up.

well as fine-tuned adjustments to answers already drawn. It definitely helps to draw!

We're most interested in the reasons people have for their answers and their arguments against other answers. Someone always reasons that the washer will continue to go outward, sometimes using the word *inertia* and sometimes simply saying that it will move in the direction it was moving—some people who've studied physics say it will continue along the tangent to the curve at the end. In almost all conversations someone connects this question to the experience of swinging on a swing and jumping off the end, usually to say that you go outward.

A Useful Subquestion: Does the Pendulum Stop at the End?

Often someone argues that the pendulum stops at the end of its swing, so letting go of the pendulum at that instant would be just like dropping it from a standstill, and it would fall straight down. Most others don't find that convincing. They question whether the pendulum actually *stops* at that instant, and this has often made for a useful digression: Does the pendulum stop at the end point? One response is that if you let go of the pendulum a little before the end point, then it is moving outward. But if you let go a little *after* it reaches the end point (so it has started on its return trip) then it should fall *inward*. As it goes outward, it moves more and more slowly, so it comes closer and closer to falling straight down. As it begins its return trip, it goes faster and faster, so letting it go makes it fall more and more inward. In our original seminar, one teacher came up with a thought experiment: If you put a pane of glass at the end of the swing, and you placed it perfectly so that the washer just touched the glass, would you expect the glass to break?

Another Clear Argument: There's a Tug-of-War at the End of the Swing

For some people, the digression to talk about whether the pendulum stops is beside the point. They reason that even if the pendulum does stop at the end point, there are competing forces, a kind of tug-of-war between the outward centrifugal force on the washer and the inward force by the string. By that reasoning, if you release the string, the outward force will win and the washer will move out, even if it had come to a stop.

These are all wonderful arguments: they are about mechanism (What pulls or pushes on the pendulum?) and coherence (How does this question relate to others?).

Try the Experiment?

Very often, some participants want to find the answer by *trying it*, which is also wonderful. In this case, the experiment is very hard to do: Almost everyone lets go of the string before the washer gets to the end point. Releasing it just at the end point takes a bit of practice, and there's no question that what people believe will happen to the pendulum affects how they carry out the test. This can be an opportunity to discuss the relationship between theory and experiment, in particular to challenge the idea that experiments always come before theory.

The Physicist's Answer

If you don't want to know what a physicist would say, then you should stop reading now! The physicist's answer is that the washer will fall straight down, for the reasons many participants give: the washer stops at the end of its swing, and letting it go at that point is like dropping it from a standstill. The problem with the tug-of-war argument is that the outward force is the washer pulling *on the string*; it's not a force *on the washer*. The only forces on the washer are by the earth (the force of gravity downward) and by the string inward. If you let go of the string, the force by the string on the washer disappears, leaving only the downward force of gravity.

Our purpose in discussing the question, however, is not to teach physics—if it were, we'd spend much more time on this and related questions. Rather, it is to help participants think through possibilities for what students might say. Having this conversation ahead of time seems to help participants appreciate what the students do and do not accomplish in their discussion.

The Discussion to Line 79

Opening the Conversation

We always begin with an open question—"What do you think? How are things going?" or "If you were Mary, what might you be pleased to see, and what aspects of the students' thinking might you want to address?"

Typically the first comments are about the students' enthusiasm and engagement, how they are talking to each other, listening, and responding. Often the elementary school students' interest and willingness to discuss the question compare favorably with the workshop participants', and we commonly hear people say, "They're doing better than we did." Some raise the concern that not all the students seem to be involved in the discussion; sometimes others in the workshop disagree.

Pressing for Specificity and Valuing Differences of Interpretation

It's probably a good idea to let the group get going in the conversation, to let everyone break the ice, but at some point it's important to press for specificity: "Can you point to a particular example or two of what you mean?" In some workshops, people start doing that without any prompting, usually when there are disagreements.

One disagreement may be over whether enough students are participating, and then some people identify students with their heads turned in the video while others identify students who seem to be following the discussion. Mathew plays a strong role here and later in the discussion, and participants often have different views of that. Some comment on his good-spirited comfort in changing his mind (line 58), to "disagree" with what he had just said, without any apparent embarrassment. Some are concerned that he dominates the discussion to the expense of others' involvement (lines 68–69).

Another disagreement often arises over whether the students really are listening and responding to each other. Some point to examples of responses, such as when Shadawn (line 32) refers to what others have said, when Mathew disagrees with Ike (line 52), and the exchange between Ashley and Mathew (lines 65–74). Others interpret those examples differently, thinking that the students are only agreeing or disagreeing without giving reasons to support their ideas. Most think Chris (lines 11 and 16) only stated his answer; many think that's all Ike did too (lines 40 and 49).

Emphasizing the Substance of Students' Reasoning

In most groups, the conversation tends to focus on general aspects of students' participation. People want to talk about whether Mathew (lines 68–69) was appropriately respectful in his disagreement with Ashley (lines 65–67), for example, which is certainly a valid and important topic. Some want to talk about whether other students are listening.

Typically, however, participants need our prompting to explicitly consider the substance of students' thinking, which is our main purpose. What did Ashley *mean* by "the opposite way"? Was she thinking of what happens at the end point, that the pendulum would begin to swing inward? What did Mathew think she meant?

That's our main purpose with these case studies: for people to get practice at understanding the substance in students' thinking. So we try to facilitate or prompt for that, to focus, for example, on what Shadawn (lines 33–34) meant in saying, "I think it's like the air, that when it goes, I think, like, the air or whatever, is pushing it." Was that an example of an argument to support her agreement with Chris? Is it a plausible, tangible idea? What did Rebecca (lines 36–37) then mean by "the force of swinging it"? Was that consistent with Shadawn's thinking? Is it clear what Mathew was thinking that led him to change his mind (line 58), from his explanation (lines 60–61)?

Of course, it is neither necessary nor possible in the time of a workshop to analyze every line of the transcript. It is, however, important to linger on some of them at this level of specificity and try to interpret student thinking.

Moving from Interpretations to Ideas for Instruction

Our interpretations can then support conversations about the menu of possibilities (see Chapter 3) for the teacher. Given any interpretation, what are some options for how a teacher might respond, in this moment for these students?

People who see the students as not really supporting their answers with arguments might suggest asking them to do that ("Can you explain why you think so?"), or they might suggest prompting them with ideas ("Could this be like swinging on a swing?"). Others who agree with the interpretation that students are not supporting their answers suggest simply letting the discussion continue, thinking it has been only a few minutes and students may begin to do that on their own. Some want to ask Ashley to say more about her "opposite" reasoning, on the view that she hasn't articulated it fully; some want to ask Rebecca or Mathew or Shadawn to say more about his or her idea that a "force" pushes on the pendulum.

What people say reflects not only their sense of the students' thinking but also their sense of what science is about. Some want to turn the discussion back to the idea of gravity, partly with the sense that students should be encouraged to use technical vocabulary. Others see the same use as something to avoid, interpreting the students as using terminology without a solid sense of its meaning.

Talking About Criticisms of the Teacher

This case is no exception to the general tendency we described for people to evaluate the teacher. Very often, people have admiring comments, but they tend to make them and be done. (That might be because much of what they tend to note is the attention Mary is showing to student thinking, so it is natural to try to return to doing that.) The criticisms take more time and tend to rouse more passion—in a few workshops some people have taken a tone of indignation. The challenge in facilitating is then twofold: to defuse the tension and to shift attention back to the students' thinking.

One typical criticism is that students "have no background knowledge" from which to think about the question, so the lesson should begin with experimentation. Our approach in facilitating conversations has been to try to focus back on the student thinking as it appears in this case, to ask again, "Is there something in this data to support that interpretation?" Most of the time, this worry has come from a general sense of what would *constitute* background knowledge, and we have sometimes taken the opportunity to digress on that topic: Would experience on playground swings count as background knowledge? For some, this instance presses on their views of science and the extent to which it connects to everyday knowledge and experience.

Another typical criticism is that the question is unclear. Some people argue that the question does not specify that the pendulum has stopped at the top of the swing. This concern does have support in the data: students speak of the pendulum as moving when it is released. From our perspective, the question of whether the pendulum is motionless "at its highest point" (as the teacher identified the moment) is part of the physics students should have the opportunity to consider. Again, however, our approach in facilitating the conversation is to keep the focus on the students, here to frame the concern as reflecting an interpretation of the data: the students seem to be thinking of the pendulum as in motion at that extreme point of its swing. The suggestion to clarify the question may then be one among a menu of possibilities, given that interpretation, with advantages and disadvantages we can discuss. It could help the students arrive at the correct answer, but what if the clarification leads to confusion—maybe students will think it means the teacher will do something special to make the pendulum stop at that moment.

There may also be an opportunity here to discuss what teachers should hope student inquiry to achieve: Is the core objective that they arrive at the correct answer? Many people who would say no in the abstract find it is difficult to tolerate even a few minutes of student inquiry (so far the students have talked for about four minutes) that isn't converging on the right answer. Depending on the time constraints, productive digressions might touch on participants' views of learning, knowledge, and reasoning in science. One strategy is to have participants articulate and identify the differences in their views of learning and then see how they play out in the next segment of the video.

 ### The Discussion to Line 150

Again, we open the conversation with a general question to ask participants what there is to see in the students' reasoning, and again we ask for specific references in the video or transcript.

In almost all workshops, one of the first remarks identifies Mathew's reference (lines 114–19) to what happens when someone swings on a rope, "on movies and things and real life." Often someone points out Mathew's later reference to what happens when "you throw a bucket or a ball up in the air" (line 135). Everyone notices Victoria's answer, of course (lines 80–81 and 95–97), and some notice that Depo (lines 149–50) said the same thing at the end. Sometimes people notice and talk about all of the gesturing they see in the room.

There is almost always more of a consensus than earlier that the students were supporting their claims with arguments and reasoning. Pressed to identify examples, people usually point to Mathew's references to everyday thinking and Shadawn's arguments (lines 107–11) that the washer's motion would depend on how fast it was moving. And there is consensus that students were genuinely attending and responding to each others' ideas, such as in Vanice's disagreement with Mathew's swing comparison (lines 120–30) and Depo's disagreement with Mathew's argument (lines 138–40).

Other reactions have included renewed concerns about Mathew dominating the discussion. It has come up in many groups that he did not give Shadawn a chance to reply to Mary's question (lines 112–13), and we have sometimes replayed that portion of the videotape to determine whether or not Shadawn had time to respond. Some participants comment that still more students had joined the discussion—Victoria, Amber, Vanice, Brandon, Grace, Jeff, and Depo all contributed in this segment, along with Mathew and Shadawn.

If no one else does, we prompt for specific thinking about the substance of two exchanges, first between Vanice and Mathew and then between Grace and Mathew.

Vanice's Disagreement with Mathew (Lines 120–30)

What precisely was Vanice's objection to Mathew's comparison? People generally come to the interpretation that she was objecting to a difference in the situations. The washer is tied to the string, so when the teacher lets go of the string, it will fall with the washer. But with a rope swing, the swinger lets go of the rope and falls without it. Is that a reasonable concern? Does it reflect attention to mechanisms and/or consistency? And did Mathew understand what Vanice was arguing?

Grace's Argument and Mathew's Disagreement (Lines 131–37)

Grace (lines 131–33) said that she agreed with Chris rather than Ike because "it can't really go up more because, like, gravity doesn't go up." Mathew disagreed with the idea he heard (and Grace made a funny face, which often draws laughter—it's a nice clue to how closely the workshop participants are paying attention!). He argued that objects can go up even though gravity is pushing down, with examples of throwing a bucket or a ball (line 135). Depo (lines 138–40) then disagreed with what he thought Mathew was saying.

Some participants see Grace as having argued something different, noting that she said, "it can't really go up more." It is possible, they say, that Grace was thinking of the washer as stopped at the highest point, so that in order to go up *more*, there would have to be a force upward on it, but the string would be gone and "gravity doesn't go up." Others disagree with that interpretation, because Grace also said that the washer would go outward, so she must have thought it would still be moving.

Moving from Interpretations to Menus of Possibilities

Interpretations of student reasoning lead to many possible instructional moves. One might be to find out more about Vanice's reasoning about how the string is connected, on the interpretation that there's a sense of mechanism behind her argument. Another might be to ask Grace to elaborate, on the sense that she hadn't fully articulated her reasoning. Or to ask Victoria to explain, because she hadn't yet given a reason to support her answer. Amber (lines 88–91) said she agreed with Victoria, but that was before Victoria had drawn a straight line down; might the teacher ask Amber if that's what she meant too? With several instances of students apparently misunderstanding each other, maybe it would be good to start summarizing and clarifying the different arguments on the board. Or perhaps the teacher could ask Mathew to listen more carefully to Vanice or to Grace. Or, since many students seemed to have something to say, might she simply let the discussion continue?

The teacher's role in this segment was limited to calling on Victoria, Amber (lines 82–83), and then Jeff (line 101); clarifying what Victoria had drawn (lines 92–93)—Victoria had said "it is, like, gonna kind of curve and then come straight back down" (line 81) but she drew only a straight, vertical line; and prompting Shadawn to support her reasoning (lines 112–13). It is tempting to see any of the many other possibilities as lost opportunities. In our view, such criticisms can be dispiriting: workshop participants would rightfully be intimidated to present their own teaching, because they'd know they'd have missed opportunities too. When things are going well, a class (or professional development workshop!) is rich with ideas and activity, a myriad of opportunities, most of which *must* be "lost." We have sometimes taken this opportunity to discuss that point.

Along the same lines, the segment displays the trade-offs between breadth of student participation and serious attention to particular students' arguments. Many of the instructional possibilities are to follow up on what an individual student has said, to clarify it, help identify the arguments, help the student become more articulate. But to focus on any one student's reasoning means giving less time to others'.

Returning to Earlier Concerns

In some workshops we have returned to concerns about background knowledge and the correct answer. There is now evidence that at least some of the students had relevant background knowledge for thinking about the pendulum.

As for the objective that they get the right answer, Victoria and Depo had both given it. Would that satisfy the objective? Some people will note that so far the students haven't articulated an argument to support the answer; often people recognize one in Shadawn's reasoning (lines 107–11) that the washer would fall straight down if it was moving slowly. (What did she mean by "do all the curves and stuff"? Was she referring to the answers Chris and Ike had drawn?)*

The Rest of the Snippet

In most workshops, we don't have time to watch the video past line 150, partly because we generally spend ten to fifteen minutes beforehand talking about the question itself and partly because these first segments are so rich with student thinking. Of course there's more to see in the rest of the snippet, which continues for another twenty minutes. But please don't rush to get to it: to promote close attention to students' thinking, it is better to be thorough and careful than it is to cover the whole discussion.

In some seminars we have talked about this case over two meetings, and we've asked participants to watch the video and read the teacher's case study in preparation for the second session. There are several possible topics to take up in a workshop and to suggest participants consider as they watch the video.

---

* It may be worth noting that Shadawn's reasoning here is not only sensible but also correct: if the washer is released at an instant when it is not moving, it will fall straight down; if it is released at an instant when it is moving quickly, it will move in an arc.

The Students' Use of Everyday Thinking and Experience

There are more references to situations from everyday thinking and experience, including Victoria's point about what happened when she swung and dropped her keys (lines 157–59), Grace's idea about riding on a swing (334–36), and Brandon's thought about a very heavy box hanging from a crane (364–73). In each case it might be useful to reflect on what the student was doing with the comparison. Was the student using it as evidence, to say "I've seen this happen"? Or was the student using it as an analogy to make a point, as a situation in which the causes and effects seemed clearer?

The Students' Reasoning About Causal Factors and Mechanisms

There was further talk about how the answer should depend on the speed, as Depo summarized (lines 303–7). Ike said the answer depended on how the teacher swung the pendulum (lines 184–85), an idea Victoria tried to clarify (lines 192–96). Brandon had the idea that it would depend on height (lines 270–75).

There are several interesting lines of reasoning about weight. Ashley talked about how "the weight of the washer and the gravity [are] both pulling" (lines 203 and 330–32). Aaron compared the washer to a pencil (lines 238–42) and was reluctant to give one answer to the question (line 247). Later still, he suggested that Brandon's very heavy box "might not be able to swing very far because of the weight" (lines 400–403). Amber disagreed, giving the example of a rocket (lines 404–15), which can be very heavy but still move quickly. By "gravity doesn't pull them down," did Amber mean that gravity didn't make them *move* down or that gravity didn't pull on them? Victoria and Chris (lines 416–28) then went on to talk about how rockets might "have something, like, that prevents them to go down with gravity" (lines 417–18); what did they mean?

Jeff came back to the idea he mentioned earlier (lines 102–3) that the washer would fall straight down because it was "heavier than the string" (lines 210–11), which Brandon came back to later and expressed nicely: "It's too heavy for the string to do anything, so it just gonna drop straight down" (lines 328–29). Might this relate to Vanice's earlier comment (lines 120–21) that the washer would be attached to the string as it fell?

With respect to any of these examples, it could be useful to have participants try to explicate the student's thinking—can we articulate it more clearly?—and talk about its plausibility. Were the students building from their common sense of mechanisms, of tangible, meaningful ideas and experience? Were they trying to identify and resolve inconsistencies?

The Question of Whether the Washer Stops at the Top

Ashley talked about swinging the washer to the right, saying if you cut it "before it goes down" it will "shift to the left" (lines 225–28 and 230–33). Here she may have been building on her idea from earlier of the pendulum swinging the "opposite way." What is the idea she was trying to express? Was she thinking about the instant at the end, when the washer must change directions from swinging out to swinging back in? If so, she was making an important point other students might not have not considered, one that might shed light on the question of whether the pendulum stopped.

Vanice raised the question explicitly of whether the pendulum stopped (lines 253–54 and 256–57). Was she asking whether the teacher would release it just at the moment at the top when it stopped? Or was she asking whether the teacher would be swinging it in a special way to make the washer stop?

The End of the Class

The topic always comes up in workshops of what the teacher did at the end, whether or not we show the rest of the snippet: Did the students try the experiment? Did they see the answer? Our strategy in workshops is, as always, to try to keep the focus on the students' thinking rather than on the instructional approach. Mary discusses the end of the class briefly in her case study.

During the class, Victoria's idea that the washer would drop straight down came up several more times. And at the end of the discussion on the DVD, Jeff checked in again on the question, asking if the teacher was going to cut it "when it's about to come down?" (lines 436–37) and going on to argue that "it will come straight down because it's about to come down" (lines 441–42). Mathew then decided to agree, saying that what the washer would do would depend on the timing of its release: if the teacher cut it "as soon as it gets back" (line 444), it would drop straight down, "but if you do it exactly when it comes up, then [it] is gonna fly," because "it still has the force" (lines 457–58). Shadawn said she disagreed (lines 449–53), and Victoria commented that Mathew was now agreeing with what she originally said (lines 447–48). What should we interpret as these students' understanding of the question? Did they arrive at a correct understanding?

Interpretations and the Menu of Possibilities

All of these moments from the class allow multiple interpretations of the students' thinking, and these different interpretations support multiple possibilities for teacher response. Seeing students as struggling to articulate their ideas, the teacher might focus on that as an instructional objective, possibly by having students write, possibly with prompts. Interpreting students as losing track of the question, the teacher might remind them of it or choose to allow the digression. Possibilities that come up in conversation include, again, summarizing arguments on the board, on the sense that the students could use help keeping track (or to support Depo in his attempts, at lines 303–7 and 432–35). Many physics teachers see particular conceptual difficulties in the idea that the weight of the washer is a distinct phenomenon from gravity, and in students' thinking of force as the same as motion. They have various ideas for addressing them, generally as long-term objectives.

In some seminars, it works to have these possibilities come up through-out, moving fluidly back and forth between interpreting student thinking and considering a menu of instructional possibilities. In other seminars, ideas about teaching take over and become detached from this case. Then we have to intervene to get participants to focus again on attending to the students' reasoning.

Mary's Case Study

We have always shown and talked about the video, at least through line 150 of the transcript, before people read Mary's case study. Showing the video first, we think, helps the case seem more real. It also gives workshop partici-pants the chance to form their own ideas about what is happening. Often their ideas align with Mary's.

Opening the Conversation

We begin conversations again with an open-ended question, such as "So what did you find interesting?" Typically, some people comment about this being Mary's and her students' first time having a science talk. Those who are familiar with Reading Recovery may talk about its compatible emphasis on teachers' attention to and interpretation of student thinking. For some, these values are familiar, and, like Mary, they find it new and interesting to think they might also apply to science.

"My Students Would Never Do This"

Very often, too, people want to hear more about the class—the governing rules, what kinds of conversations student had had before this. For many, it seems mystery that students could engage in a discussion like this. Often people feel that this sort of conversation could not happen in their classes, and they want to understand what was special about these circumstances. Mary mentions that twelve students were designated as gifted, but this was not a magnet program, and there was not a special selection for the students. We try to deflect the idea that this was an anomaly, mostly by simply telling participants that in the project we've grown accustomed to being impressed by children. It is, we argue, a matter of giving children opportunities to express themselves and, when they do, taking what they have to say seri-ously. This is an opportunity to turn the conversation toward that aspect of Mary's practice.

Promoting Attention to Mary's Interpretations

We often prompt participants to pay attention to the teacher's interpretations in discussing the case. Rather than focus first on what she *does* as the teacher, we ask participants to focus on what she *sees*. What are the interpretations she makes that motivate her to respond? One strategy for structuring the conver-sation is to ask participants to identify moments in the case study when Mary recounts her interpretation of students' reasoning and how she responded to that interpretation.

Examples might include her sense that Ike had a new idea he needed to draw on the board, her figuring out what Victoria meant in saying "it curves" but then drawing a straight line down on the board, and her noticing that students were not yet mentioning examples from everyday experience. In some cases, such as hearing Mathew and others starting to refer to everyday experiences, or seeing Vanice's objection to Mathew's rope-swing argument as showing perceptive attention to detail, Mary made the assessment quietly, allowing the discussion to continue.

Discussing the "Should Haves"

It is both inevitable and desirable that participants will take issue with some of Mary's actions (or inactions). Some, for example, argue that she should have focused more on the terms *force* and *gravity* and their definitions; summarized Shadawn's and other arguments on the blackboard; helped Grace clarify what she meant in saying that "it can't really go up more because gravity doesn't go up"; and (more than any other) had the students try the experiment sooner.

With respect to any of these, we try to encourage two things. First, we press for appropriate respect and humility from participants, not only in this case but when criticizing their colleagues' teaching or their own, if they take up practices of presenting and discussing snippets. In almost all cases, "should have" is naïve, but "might have" is thoroughly appropriate.

Second, we press for explicit conversation about what interpretation of students' reasoning motivates the "might have" possibilities for teacher moves. In some cases, the move is motivated more by general assumptions about what should happen in science instruction than by particular interpretations of student thinking in the case. That often characterizes the interest in the terminology and in conducting the experiment. Depending on the workshop, we might take the opportunity to examine those assumptions. For example, what are general considerations for and against pressing students with respect to terminology? (There is reason to believe from research on learning in high school and college that students come to see science as about using the right words rather than about meaningful thought.) Or, what are general considerations about when to try an experiment? (In some situations experiments can actually curtail productive inquiry, if students consider the experiment to have answered the question and believe that answer to be the end of the topic.)

Other criticisms are based on specific attention to students' meaning. Participants' suggestions that Mary might have made a list of students' arguments on the board generally come from their seeing the discussion as fragments; their suggestion that Mary might have followed up on Grace's reasoning comes from an interpretation that she did not mean what Mathew thought she meant.

Finally, participants often find matters to discuss in Mary's descriptions of what she was looking for in the students' reasoning: whether they were expressing themselves clearly, being tangible, listening to each other, and thinking about how to address opposing ideas. These are often useful topics, focused on details of Mary's interpretations. We try to use particular instances in the case study to help illustrate and clarify, but depending on the workshop, this may be another place to digress into more general conversation about science teaching.

Often some participants are concerned that Mary should *also* be paying attention to, for example, whether Mathew is appropriately considerate of other students. She *was* paying attention to those things; she just hasn't *written* about them. It's not possible to write about everything, and our focus in these cases is on the substance of students' thinking.

CHAPTER 5

First Graders Discuss Dropping a Book and a Piece of Paper

Jamie Mikeska was about to start her second year of teaching when she joined the project in the project's second summer. She wrote this case study about a pair of classes she taught that November. It has video in two parts: Jamie videotaped one discussion herself, on one day, and project staff visited and taped the continuation of that lesson on the next day. We've found people have an easy time following and interpreting the children's thinking in the first part of the video, which makes this a good first case to watch. It gets harder to follow their thinking in the second part, as you'll see.

Jamie's case study begins on the next page. As always, we recommend watching at least some of the video before you read the teacher's account, because that gives you the chance to discover aspects of the children's thinking for yourself.

Suggestions for First Viewing

Be sure to print the transcripts from the DVD-ROM for "Falling Objects Day 1 and Day 2." You can use them to follow the conversation and take notes, and you can use them for study and reference afterward.

Day One
The day one video clip beginning with "I have a book and a piece of paper" is about five minutes long. We generally watch it all at once. (The conversation went on for another twelve minutes, but we stopped it there to save space for the next day's video.) You could watch this clip without reading anything first. It opens with Jamie asking a question, and we've found it is fairly easy for people to interpret the students' thinking. Or you could read Jamie's introduction first, to get a bit of background about her class; that would also give you a sense of how she arrived at this question and what she was hoping to accomplish.

Remember to focus on the substance of the children's thinking: What do you notice? What might be the beginnings of scientific inquiry? If you have thoughts about what the teacher should or shouldn't have done, try to articulate the interpretations of children's thinking that those ideas reflect: What about the children's inquiry gives you that sense of how the teacher should act? Can you be specific?

 Day Two

For the day two clip, it will help to read the beginning of the case study first, even if you've watched day one. If you want to go into the video fresh, without Jamie's take on it, stop when you get to the heading "What Happened?!" on page 75.

We recommend watching the video in segments. Watching a smaller amount at a time makes for closer examination of what children say; there's so much even in a five-minute clip. Here are some good stopping points for reflection and conversation.

1. *When Henry says, "No I didn't" (line 71 on the transcript), three minutes into the snippet.* What do you notice about the students' reasoning? How does this segment compare with and contrast to what happened on day one? It is tempting focus on what the students aren't doing; be sure also to think about what they *are* doing.
2. *When several children start jumping (line 169 on the transcript), almost seven minutes into the snippet.* Once in a while we skip this segment in workshops, along with the next, saving time for what comes later.
3. *Skip to DVD Scene 4, and then stop after Jamie sends the students into groups (line 288 in the transcript), about eleven minutes into the snippet.*
4. *After Diamond's comment about balling up the paper (lines 355–360), at fifteen minutes.*

That leaves six minutes of video until the end of the discussion, which we hardly ever get to in our workshops, and never when we've also watched the clip from day one.

When you've had enough of a chance to watch the video and have some ideas about what happened, read Jamie's account of the class. And if you'd like still more, or if you're planning to lead a conversation, read the facilitators' notes on page 83.

■ ■ ■

Falling Objects

Jamie Mikeska, Montgomery County Public Schools, Maryland

My first-grade students were already experienced at making observations. In our classroom, nothing seemed to escape their attention. When the classroom furniture was rearranged, they noticed. When our independent reading time began a few minutes late, they noticed. When something did not go as planned, they noticed. They were also experienced at using their observations to draw conclusions. When they saw it raining outside, they knew not to get their coats for recess because they knew it would be indoors. When they saw me getting out my writing journal, they knew that it was time for writers' workshop.

Applying their everyday observations to observations in science inquiry seemed to flow naturally. *Making observations* and *drawing conclusions* are two essential components of the county's science curriculum, but my students were already doing both. I also knew they would not be challenged by the county objectives that they be able to "give examples to demonstrate that things fall to the ground unless something holds them up" and to "describe the different ways that objects move (e.g., straight, round and round, fast and slow)" (MCPS 2001).

I wanted them to go further than these basic objectives, to use their abilities to make observations and draw conclusions to delve deeper by asking *why* questions. I consulted the science specialist for ideas, and he suggested a classic experiment: drop a flat piece of paper and a book at the same time from the same height, and then drop a crumpled piece of paper and a book at the same time from the same height, and compare the difference in results. That seemed perfect.

The students were comfortable both working in small groups with hands-on materials and sharing their thoughts and results in a whole-class setting, so I planned some of each for the activity. I assumed that they would first observe the book hitting the ground before the flat piece of paper and then the crumpled paper and the book hitting the ground at the same time. I could then ask them to explain the difference in results, and I was interested to see how they would do. But the lesson did not work out quite as I had planned.

Predictions: The Book and the Flat Paper

Before the first experiment, I held a book and a sheet of paper out at the same height and posed the question to the whole class, "What will happen when you drop a book and a piece of paper at the same time?" Almost immediately, I heard a chorus of voices saying that the book would fall down first and starting to give ideas of mechanisms for why that would happen.

Ebony explained that "the book is going to fall down first because" it had "more strength" than the paper. He said, "If you put it on, um, the things that you weigh yourself on, the paper is going to weigh nothing but the book is gonna weigh like one or three pounds." I asked the class to talk about Ebony's idea, and other students agreed with it.

Allison said she thought "the book has more force." Autumn explicitly related her ideas to her previous experiences. She said that she knew that the book would hit the ground before the piece of paper because "that's what really happened when I did it at my dad's house. I had a Pinocchio book and I had a piece of paper and I dropped them both. The book fell down first."

One of the last ideas that surfaced in our whole-group conversation was that the book would fall first because the paper would "float." The students talked about how the paper would float from side to side, slowly making its way to the floor. They even showed the flight of the paper using their hands. They explained that "the wind" would make the paper move that way. One student said the air was "blowing it."

The First Experiment: The Book and the Flat Paper

The next day we began by watching our videotaped discussion from the previous day. I wanted the students to understand and process each other's ideas for why the book would fall faster than the paper.

We watched the video and stopped each time someone shared a *why* idea. I wrote the ideas on chart paper for the students to use as a reference, possibly in their future conversations with one another.

We went on to other things, but later in the day we came back to our list of *why* ideas. The students were brimming with excitement; they couldn't wait to have a chance to do the experiment on their own. I began by showing the students what I meant by the word *drop*. Because ten out of the sixteen students in our class spoke a different language at home—Spanish, Russian, or Amharic—I wanted to make sure everyone had a firm understanding of the question.

As I modeled how to drop the objects, I explained, "You don't throw it down. You just let it go. When you drop it, you just let go of the book. Your arm barely moves." A few students actually dropped the book for practice and we were on our way. I divided the students into small groups of two to three students per group. Each group went to an area of the classroom to work.

The classroom exploded with the noise of students talking and books hitting the floor. Some groups took turns letting each person drop the book and

---

What will happen when you drop a book and a flat piece of paper at the same time?

The book will fall first.

Why?

1.   The book has more force.

2.   The book is stronger than paper.

3.   That's what happened when she did it at her dad's house.

4.   The paper will float.

The Ideas We Recorded.

piece of paper while others had one person hold the piece of paper and someone else hold the book. Then, the partners would count to three and let them go. I heard, "Look, Ms. Mikeska! The book hit the ground. It went fast." Another student exclaimed, "Hey, the paper goes slow. It goes side to side." My role during this time was only to observe. I watched some of the groups do their experiments, but I could not watch them all.

After a few minutes, I gathered them back together so they could communicate their observations to each other. Then, I thought, we could focus again on the reasons that the book hit the ground first.

What Happened?

I asked the children to tell me what happened in their experiments.

Teacher: What happened when you dropped the book and piece of paper at the same time, at the same height? What happened? OK, Ebony, why don't you go ahead and begin.

Ebony: To me, first, the paper fell first.

Several Students: No way! No! Whoa! The book fell first.

Ebony: No, to me the paper fell first.

Student: It fell at the same time.

Ebony: No, the book, um . . . the paper fell . . . the paper fell first to me!

Henok: Yeah, but not to me!

Jorge: Yeah, but did the book fall first, just like the paper?

Ebony: No, the papers fell first. No, the paper—to *me*.

Allison: To Ebony—to Ebony, the paper fell first.

Brianna: And to all of us the book might of fell first to us.

My mouth dropped open (or at least it felt like it) at about the same time that the students were shouting, "No!" How could it be that Ebony observed the paper hitting the ground first? Did he drop the piece of paper before he dropped the book? Did he drop them at the same time but from different heights? As the discussion continued, I hoped that some students would realize that Ebony must have done something differently in order to cause this shocking result. I was in for another surprise, though, from Rachel.

Rachel: With me and Julio, twice the book and the paper tied—twice.

Ebony: —went at the same time.

Rachel: They both tied twice.

Allison: What do you mean, "tied"?

Julio: They both fell down at the same time.

Allison: Same with me, same with me, same with me, same with me!

Teacher: So, what I'm hearing is we have one person that said when he did it, that when he dropped [them] at the same time, from the same height—

Brianna: The paper fell first.

Teacher: —that the paper fell first. Now do you mean it hit the ground first, or it just started to fall first?

Ebony: It hit the ground first.

Teacher: So, you're saying the paper hit the ground first, and then the book hit the ground. [*Ebony nods his head in affirmation.*] Then we have two other friends who are saying that the book and the paper hit the ground at the same time.

Rachel: Twice!

Other Students: Yeah! Twice!

Allison did a nice job when asking Rachel, "What do you mean, 'tied'?" That could mean that the book and the flat piece of paper fell down side by side until they hit the ground, but it could also mean that they only hit the ground at the same time. Maybe one started falling before the other, and the second "caught up"? But Julio said they "both fell down at the same time."

At this point, I was almost scared to listen to *other* students describe *their* observations! Maybe someone would say the book floated upward! I asked, "What happened with it when the rest of you did it?" and to my relief, students started to say "the book fell first." Diamond then complained about how Henry was "putting the paper" to make it fall first.

Diamond: Um, the paper, when we . . . the book and the paper . . . the book fell first . . . but Henry keep putting the paper so it could fall first.

Teacher: How was Henry putting the paper so it could fall first?

Brianna: He's dropping the paper first and then the book.

Diamond: He did like this. [*Raises and lowers arm.*]

Henry: No!

Diamond: Uh-hm. And I told you to stop.

Henry: I didn't do that.

The day before and earlier this day, the students had all predicted the book would fall first, and that's what I was expecting they would all see. Instead, the students had observed three different results.

I had planned to move on to the question of the crumpled paper, but now I wanted them to explain the different results. Diamond had already begun to give one possible reason: Henry was "putting the paper so it could fall first." Looking back, I realize I could have pushed more on her idea as a way to start the class thinking about the new question I had for them, but I just asked it: "How could it be that we all . . . got different results when we did the same thing?" I wanted the students somehow to realize that when you do science, in order to draw any reliable or reasonable conclusions, you must be able to replicate results. This was a different focus than I had originally intended, but I was concerned that some of the students were thinking that it was OK for different people to have different results.

At first the students just repeated their observations. They were not answering the question or providing any reason as to why we might get different results. At least one student, Allison, shrugged, showing that she understood my question but didn't have an answer. I persisted in my questioning until Rachel proposed an explanation.

Rachel: Forces of gravity.

Allison: Yeah!

Diamond: What are forces of gravity?

Rachel: Gravity is what—

Allison: Gravity—you know how when we jump, we always land back on the ground.

Rachel: Exactly. It's what keeps us down on the ground.

Ebony: No gravity is when you're, like, in space and you can never ever really fall down.

Allison: Gravity—see how I jump. [*Jumps.*] I'm just landing at the same place on the ground . . . because gravity . . . is just pulling me down.

Rachel introduced a science term, *forces of gravity*, that I expect few of the students had heard before. Diamond asked a good question, "What are forces of gravity?" which prompted Allison to give meaning to the term through a familiar experience, jumping up in the air and coming back down to the ground, and Rachel agreed. Even so, they did not seem to know fully what this word *gravity* meant, and the terminology did not do anything to answer the question of why different groups got different results. I probed on.

Teacher: OK, so what you're saying . . . you're saying the force of gravity—

Rachel: Is pulling it down at different times.

Teacher: So, you're saying the force of gravity is pulling the book down at a different time than the paper.

Rachel: Yeah, probably. And, sometimes it's pulling it down at the same time, or pulling the paper down before the book, and then the book's pulling it down before the paper. Gravity's pulling the book down before the paper.

It sounded as though Rachel thought the force of gravity was a variable that acted according to its own whims, and my sense at the time was that this did not seem like a reasonable explanation.

But conversations later with other teachers gave me more to think about. Maybe Rachel was being more reasonable than I thought. She might have been thinking about how other things in nature have a varying effect, such as the wind: If you threw a ball from the same spot, in the same direction, at the same speed over and over again, the ball would not land in the same exact place every time. The wind might change speed or direction, causing the ball to take a slightly different course each time. Of course, the ball would probably land in the same place a few times, just like the book and the paper hit the ground at the same time more than once. Thinking about her idea in relation to this example made sense to me, so maybe it could be a reasonable explanation for these first graders. Gravity changes just like wind changes. Earlier, when I had asked them directly to resolve gravity's inconsistent effect, it is a possibility that they did not answer because they did not see the inconsistency.

At this point, the students were out of ideas—or perhaps out of steam. Or maybe they were confused by my asking them about an inconsistency that they did not see. In any case, I did not feel many of the students were comfortable with this explanation. The mechanisms they had talked about during their predictions, that the book would be pulled harder to the ground and that the air would push up on the paper, still seemed the most tangible explanations they had, and I wanted to get back to them. Since it looked like the class was tiring out, I decided to drop the book and flat piece of paper in front of the whole group, hoping to generate more discussion. It did not work. Instead, the students became a little silly, talking and giggling about how the book and the gravity were getting "tired."

It was time to move on. The next step was to introduce a new situation to these students: dropping a crumpled piece of paper and a book at the same time from the same height.

Predictions: The Book and the Crumpled Paper

I began by crumpling the piece of paper in my hand into a small wrinkled ball, to show the students what they would be doing. I was interested in what they would predict would happen in this situation. After the last conversation, I thought they might predict "gravity" would cause three different results again. But the word *gravity* never even entered the conversation! It seemed as though the students had forgotten what was discussed just a few minutes earlier— maybe it just hadn't been meaningful to them.

Teacher: What do you think will happen?

Jorge: The book, um, the book . . . and the paper will fall first.

Brianna: At the same time.

Jorge: Yeah.

Diamond: The book and the paper will fall at the same time.

Jorge: When you crush it. When you crush it like this and it will fall down first. Remember, I did that while at home.

Diamond: It gonna fall at the same time.

Jorge: I used to do that all the time.

Jorge explained that he thought they would hit at the same time because he had tried this at home. But nobody else had. Since no one was offering a different prediction, I changed the question from "What will happen?" to "Why do you think that?"

Teacher: Why do you think that? That now the book and the paper will fall at the same time.

Allison: Because now the pap—

Diamond: When you ball it up, it's almost like a ball—

Allison: —now, now—

Diamond: —and the ball falls first, too, at the same time as the book.

Allison: Now, now the paper has a little more weight because it's all crumpled.

Jorge: Yeah. Like crumpled. Like—

Allison: So it has a little more weight.

Allison's idea that a crumpled piece of paper has more weight than a flat piece of paper sounded like one of the old Piagetian results of preoperational children not conserving matter. I wondered if she really meant that it would be heavier; if I'd had a scale in the room, we could have tried weighing it.

We had spent a lot of time sitting in the same circle and talking, and the students were all predicting the same thing, so I decided to let the conversation stop and have them go try the experiment.

The Second Experiment: The Book and the Crumpled Paper
As I circulated around the classroom this time, I heard a lot of students saying the book and the crumpled piece of paper hit the floor at the same time. Once more, we formed into our circle and I asked them to say what had happened. This time, everyone agreed on the results.

Teacher: So what happened, guys, when you dropped the book and the crumpled piece of paper? What happened this time? Brianna. Brianna will go ahead and begin. Remember, we just want to talk one at a time.

Brianna: They will fall at the same time. 'Cause they both got the same strength together.

Teacher: So Brianna has an idea. She said, well, she found out that they fell at the same time because they both have the same strength together?

Brianna: Strength together, yeah.

Teacher: Before, we said that the book had more strength than the piece of paper. How could it be that now they have the same strength? We are going to talk about Brianna's idea and then we'll go to the next one. How could it be that now they have the same strength?

Rachel: I can answer that.

Teacher: OK.

Rachel: Um. Now that it's crumpled up, there's—the piece of paper has more strength because—

Brianna: And the book too. Because they have [the] same strength all together.

Rachel: No, I said—the piece of paper has more strength because it's all crumpled up and it used to be, um, really light but now it's, um, it has more strength. It probably has as much strength as the book since all the, um, paper is crumpled up together.

Brianna shared an idea that the book and the crumpled paper "got the same strength together." As before, she seemed to be using the word *strength* as a synonym for the word *weight*. Rachel's response was to help clarify Brianna's reasoning: the paper had the same strength as the book because crumpled paper weighs more than flat paper. This idea had surfaced earlier in the students' prediction conversation, when Allison discussed how crumpling a piece of paper makes it weigh more. Could they be thinking that when you crumple up a piece of paper, it becomes more solid, like a ball, and that's why it seems heavier?

The idea that the paper weighed more did not sit quite right with Brianna, who took the ball and uncrumpled it. Another student asked, "Why are you doing that?" in a tone of voice that made everyone laugh, and Brianna answered that the paper was still the same size.

Brianna: It's still at the same size. It still feels [*laughter*] . . . it still feels, um . . .

Autumn: It's not heavy.

Brianna: It still feels . . . it still feels light.

Allison: My, my dad could probably throw it up to the ceiling and he wouldn't, and he wouldn't say it's light.

Brianna: It's still light.

Student: It's not heavy.

Julio: It is not.

Teacher: So are you disagreeing with Julio?

Brianna: Mm-hmm.

Teacher: So you're saying that it's, it's . . . this is still light even when it's balled up?

Brianna's reasoning that the piece of paper was still the same, even when it had been balled up, reminded me again of the Piagetian idea of conservation of matter. She seemed to be saying the paper was still the same paper, even if we changed what it looked like. We were not taking anything away from it or adding anything to it to change how light or heavy it felt. Brianna's reasoning moved the students' thinking forward. They stopped talking about the paper being stronger or heavier than it was before, and they started thinking.

Diamond: Because the piece of paper was like . . . the first time, like this, and then it balled up.

Teacher: So, you're thinking it's the change in—

Allison: In shape.

Teacher: In the shape?

Diamond: Yeah.

Teacher: And that's what caused the difference?

Allison: Because then you change the shape. At first it's a rectangle. Then, it's a sphere.

Brianna: It's a rectangle.

Teacher: OK.

Diamond: Now, it's a sphere.

Teacher: Go ahead and put it in the middle there. . . . You wanna share with the group?

Diamond: Now, because now the piece of paper can roll.

Henry: If, if the book is the same, like, heavy, and, and you go in the same time, like, like, at first at the same time the book will fall, like same time, like this. It will fall, it will fall, like that—'cause look. If you put it in front of [*talking about the flat paper*]—

Brianna: The, the book will fall first.

Henry: Because if it, if it goes like this, then it will go like that.

Diamond: I was not done what I was sayin'.

Brianna: The book. The paper—

Diamond: I was not done what I was sayin'.

Henry: —will fall last.

Brianna: Yeah. And the book will fall first.

Diamond: 'Cause the piece of paper was balled up, it don't go like this no more [*shows a rocking motion with her right arm*].

Brianna: No, yeah. It don't, yeah. It just drops, kind of like the booklet.

The students zoomed in on the change in the shape of the paper, which they felt was causing the change in the paper's movement. When the paper was flat, it floated from side to side. When it was in the shape of a ball, it dropped straight down, just like the book. It helped that the students had the materials in front of them during our whole-group discussion so they could use the materials to show each other what they meant.

The conversation came to a close with the students continuing to discuss how the paper moved when it was crumpled into a ball and how the book moved. When they started talking about the flat paper moving back and forth, I thought they might come back to the mechanism of the air pushing up on it, but they didn't. I ended by asking the children to try this out at home with two objects of their choice. My purpose was to extend what we had been doing in school to a new, yet similar, situation at home.

Afterthoughts

As I reflect on these shared experiences, something in the back of my mind keeps nagging me. How could the students make predictions that seemed very logical and reasonable one day and then return with observed phenomena that seemed to completely refute their sensible ideas the next day? How did they not question these observations? It seems as though their sense making was completely cast aside as they shared their observations with the group.

Looking back, I wonder what would have happened if I had just skipped the part where the students went and tested the flat paper against the book. The previous day's discussion showed the students already knew what was going to happen. Autumn even went so far as to say that she had done this before at her dad's house. It is possible that the first question was just not engaging enough for the students. Through discussions with colleagues, one idea surfaced that Ebony might have been trying to make the question more interesting. Maybe he was playing with the question. His initial comment of "To me, the book fell first," especially how he said this, seems to suggest that he knew his observation would be controversial.

Similarly, why do students share ideas, such as the idea of the air affecting how the paper moved, and then never mention or come back to them in subsequent conversations, even when they seem relevant to another situation? This is definitely not the first time I've seen this in conversations with young children! Thinking about the students' ideas and the flow of the conversation has me wondering how I might help them learn to keep better track of their ideas. It might have helped if I had charted the students' thinking across the different parts of the experience, in the same way that I charted their beginning ideas for why the book would hit the ground first. My purpose for the chart was to help the students focus on listening carefully to others' ideas. If I had referred back to it and made new charts as they spoke, maybe it would have made the students' thinking more visible. Could they have used it as a reference in subsequent conversations? I would like to help them think, "Do these new ideas or observations fit into what we were saying? How do we need to revise our thinking?"

■ ■ ■

Facilitators' Notes

Please see the general notes for facilitators in Chapter 3. Here we'll provide specific comments and suggestions with respect to talking about the snippet at the stopping points we suggested earlier.

In some places we'll note topics we make sure come up; other topics we consider optional, and we certainly don't recommend trying to get to all of them. We drew from many different workshops and seminars to put together these notes; none of them hit all of these topics. And, of course, any given conversation will touch on things we haven't addressed.

 ### The Snippet for Day One
It wasn't until late in developing this book that we decided to include the video from the first day. Before that, we'd started workshops on this case by giving a summary of what had happened the first day. Now we always start with at least a short clip from the first day, which does much more to give people a sense of the children's thinking.

Opening the Conversation
We ask what people see in the children's reasoning, often phrasing the question in a couple of different ways: "What do you see or hear in the children's reasoning? Is it what you expected, or does anything surprise you?"

Some people aren't used to watching six-year-olds, and they need a moment to adjust, maybe to call the children adorable. People used to six-year-olds tend to start talking about how these students were using knowledge from their experience to answer the question. Some want to talk about which students were speaking up and which were not. And some feel compelled to talk about the teacher.

It's always a matter of judgment how much to let things get going before steering the conversation toward the substance of the children's thinking. For this case study more than others, it's also a matter of judgment how long to linger on the first day's clip while still leaving enough time for the more surprising and problematic second day. Regardless, there are several general directions in which it's important to guide the discussion.

Focusing on the Substance of the Children's Thinking

If things aren't getting started easily, we might ask a more directive, specific question: "What are the children's ideas?" Every group should be able to identify the children's reasoning about "strength" and about "air"—more on that in a moment.

Usually, though, things do get started easily and quickly build up some conversational momentum. It's usually not quite in the right direction, though: Because we always use this case study as the first or second one people see, part of the task is to help everyone understand the sort of conversation we're trying to have. That always involves some work, especially to focus attention on the children's thinking rather than on what the teacher is or isn't doing.

One way to do that is to stop the conversation and refocus it. But it's often possible to reframe the conversation instead, to use what people have said about the teacher to redirect the discussion. For example, people sometimes admire how the teacher helped the children focus on Ebony's idea by asking them to repeat it (lines 53–56). That's a comment about the teacher, but it includes an implicit sense of what the children need, which we can try to make explicit: "What is it you see or expect from the children that makes you think that's a good strategy?" Maybe the answer is that children need to learn to listen to and consider each other's ideas, and we can continue from there: "Can we find some examples in this clip to get a sense of how *these* children are or aren't listening to each other?"

We've also heard the criticism that the teacher did not help the children connect their reasoning about air with their reasoning about strength. In one seminar, this was how the topic came up that the children had these different ideas; we simply pointed that out: "You're saying the children have two different ideas they aren't connecting—can we start there? What are the two ideas?" The group can look at the relevant lines of transcript for evidence of those ideas about air (lines 103–15) and about strength (e.g., lines 49–52).

We can talk about two aspects of this observation: that the children had a sense of something that would push or pull the book or paper to the floor, and they had a sense of something that would *hinder* the paper from falling. The essence of the criticism is an interpretation that these are two different mechanisms, and the children hadn't thought about whether or how they might fit together.

Having talked about these parts of the children's thinking, we could then come back to the topic of what a teacher *might* do to respond, and we take care in framing this question. "What are some items on the menu?" Participants have said that the teacher could have asked children whether the air was also blowing on the book; written the ideas on a chart and asked the children if they were the same; let the discussion continue but made a mental note of the topic; and so on. The point is not that we don't want to talk about instructional strategies; it's that we want to talk about them *based on* interpretations of the children's thinking that we've explicitly considered.

Talking About the Role of Vocabulary

One observation that has come up several times relates to the children's choice of terms: Allison used *force* (lines 21 and 32), which some people see as preferable—more scientific—than *strength*. Some note that Ebony used the word *weigh*, and they consider that another appropriate term. These interpretations can then lead to ideas for how the teacher might encourage students to use scientific terminology.

We think it is important to challenge these assessments, and typically someone does. Some people have wondered whether saying "force" was "trying for science speak," as one person put it. Someone familiar with the concepts might point out that Allison didn't mean force as physicists define it. In one group, someone questioned whether Ebony actually meant "weight" when he said "strength." We read Ebony's comment again (lines 49–52) with that possibility in mind, and we watched the video again to see how he responded when the teacher asked if he meant the book "weighs more" (lines 69–71). Was he hesitant in nodding?

Once in a while, though, no one challenges the vocabulary assessment, and so it's up to us. "There's another view of this I'd like to put on the table," a facilitator might say. (There's another good opportunity to bring this up later, after the first segment of day two, if for some reason it doesn't seem appropriate to do it here.) Thinking from a child's perspective, finding out it is better to use the word *force* than *strength* because that's the word scientists use might encourage him to think of science as all about using the right words, something we know to be a problem with older students.

To be sure, what complicates these moments in workshops is that some participants in workshops and seminars understand science that way, as about using the right words, words that may or may not have tangible meaning. In some sessions with this case, we've ended up in a digression about the role of vocabulary in science, and we offer the view we discussed in Chapter 2: the reason there are technical terms is because scientists need to express ideas precisely. If the focus is on children's inquiry, then they should be learning to express *their* ideas clearly and precisely, and that might not happen if they think they're supposed to use certain words and stay away from others.

This topic led to an exchange in one group over whether the teacher should discourage the students from using the word *force*. From our perspective, there's no reason in general either to privilege the word *force* or to discourage it; the point is to focus on meaning. So the question is what Allison meant by it, and in that group, the consensus was that she meant something very reasonable, the same or similar to what Ebony meant by *strength*.

Identifying the Beginnings of Science

In the end, we like to have a short list of children's ideas. Most of these will come up on their own in response to general prompts ("What else do you see in the children's thinking?" or "What else might be the beginnings of scientific thinking?"). If not, we might quickly list them at the end of the discussion, by way of preparation for day two:

- The book will fall more quickly because it has more strength (or force or weight) (lines 21, 27, etc.).
- The paper will move back and forth on the way down (gestures near the end of the clip, lines 94–97 in the transcript). This may be another kind of explanation for why the book will hit first: it will take a more direct route than the paper.
- The reason the paper will move back and forth is because the air will be blowing it (lines 103–15).
- Students can use their experience with dropping things to help them answer the question (lines 80–82).

There's one other thing we often point out, as a little more preparation for day two: The students began explaining at the level of mechanism without any prompt for that from the teacher. She'd asked, "What will happen?" and starting with Allison (line 21), they began on their own to give causal explanations.

 ### *The Snippet on Day Two to Line 72*

Opening the Conversation

It's easy to open this conversation, because the clip is so striking. Someone always brings up Ebony's result that the paper fell first—we like to point out that was a surprise for the teacher, too—and things proceed from there. Often, they proceed in a direction toward trying to understand how Ebony could have gotten those results. Often, too, there is talk about how the students accepted the different results as if they were all OK. So most groups start out noticing things about the students.

Ebony's Result

Ebony must have done something wrong! Sometimes that thought comes out as a question about the teaching: Was the teacher clear about how the students should conduct the experiment? Again, that implies an interpretation, and we can shift attention directly to that: Does the evidence we have available support the interpretation that Ebony was confused about what he was supposed to do?

It's surprising how often people forget it was Ebony who had been so articulate the day before about why the book would hit first; we have to remind them it's the same boy. Some people talk about the ambiguity in his claim that, *to him* the paper fell first (lines 4, 8, and so on). Did that mean that he saw the same event differently, or did that mean that when he dropped the two objects, the results came out differently?

Or, someone might suggest that Ebony thought the question was too easy. Of course the book would hit the ground first! Maybe Ebony took it as a challenge to drop them at the same time but make the paper hit first.*

The Students' Acceptance of Conflicting Results

We raise this point, if nobody else does: the students seem to be accepting different results, as if everyone is entitled to the outcome of his or her choice. Here, as in general, it is useful to look for the evidence in the data, such as in lines 18–20. That stance, people generally agree, is not good science.

This is often a moment where the emphasis of the conversation is on what the students are not doing, and then on how to get them to start. They're not being mechanistic; they're not trying to be consistent. These are useful observations, especially as they contrast with what happened on day one, when the children brought up mechanistic reasoning by themselves.

But talking only about what they're *not* doing doesn't help with understanding what they *are* doing. Why would this behavior make sense to them? One possibility people have suggested is that the discussion had become show-and-tell, a familiar activity in which everyone has her or his own thing to say. Another is that they were trying to avoid an argument.

Some may notice how Rachel was so careful to emphasize that her group found the same result twice (lines 25, 27, etc.). Why was it important to her to say *twice*? Is this something to see as the beginnings of good scientific behavior? This leads to one idea for instruction, to call attention to Rachel's point and make explicit the idea of replication, perhaps from there guiding the students to the task of figuring out what result can happen reliably. Another idea that's come up several times is to start a chart to keep track of the different results. That would show the most common result was that the book hit first and perhaps help frame the task of explaining the discrepancies.

In one group, a teacher remarked that Diamond's complaint about Henry (line 63) was an attempt to explain how the outcome could be different, that it was because he "kept putting the paper so it could fall first." The participant

---

* One way to do that is to put the piece of paper under the book, rather than next to it. Then the book lands on top of the paper.

tied this to a possibility for instruction—using Diamond's idea to get other students to try to give causal explanations for the different outcomes.

Discussing Possibilities for Instruction

Most groups will have already been talking about ideas for instruction; much of the work in facilitating is to ask what interpretations underlie the suggestions. But if a group hasn't discussed this topic yet, it might be good to ask them about it, because the next segment begins with the teacher's intervention. You might ask, "What are some options at this moment for what the teacher could do?" Sometimes we ask about the possible advantages and disadvantages for the different options. For example, some people think of asking Ebony to show the class how the paper hit first. He might have liked that, and it would certainly have helped clear things up. On the other hand, it could have set him up for embarrassment.

The specific set of possibilities here is not as important as that there are several and that they're based on some interpretation of the students' reasoning. As well, it's important that they be respectful of the teacher and the challenge of responding in the moment. We usually remind everyone that it's easier for us, watching a tape and taking our time. For Jamie, the surprises came in real time, and she had to interpret them on the fly.

 ### *The Snippet to Line 163*

Once in a while we skip this segment. That saves time and can help keep a group focused on the children, since the segment begins with the teacher's intervention. On the other hand, if a group has had a good conversation so far, it can be nice to follow through, to see what the teacher did and talk about her interpretations.

"We Did It Twice"

One disagreement we've heard is over Allison's explanation that her group "went around two times" (line 85). Was she giving that as an answer to why they got different results, or was she simply giving a more thorough reporting of her group's work? Her remarks (lines 90–96) also speak to the earlier question the group might have considered of whether Ebony had been reporting a separate trial or his perception of the same trial. His answer here, however, seems matter-of-fact: they got different answers because they "did it twice" (line 98).

"Forces of Gravity"

When Rachel mentioned "forces of gravity" (line 104), it was the second place in the case study where people have noticed technical vocabulary, and in many groups this is where we've spent time on the topic. Was she saying those words because she thought that was what the teacher wanted? Or did she think that the idea of forces of gravity could somehow account for the different results?

Some groups then go on to talk about the other students' reactions, including when Diamond asked, "What are forces of gravity?" (line 106) and the students' various answers to that. How should we interpret their jumping up and down?

We've sometimes pointed out that when the teacher started to prompt Rachel to elaborate on her idea (line 129–30) Rachel finished the sentence (132–33). Until then, Rachel had only said the term *forces of gravity*, but here she apparently realized for herself that something was missing in her explanation. One seminar spent some time on ways of understanding the idea that gravity "is pulling it down at different times" (line 128), wondering (as did the teacher) why gravity would do that. Was it significant that Rachel said "forces of gravity," making it plural? (Jamie considers a possible interpretation of this in her case study.)

Henry's Reasoning

In a few groups the question comes up of what Henry was trying to say (lines 151–60), and when it does, we play a little more of the video into the next segment, since the exchange continues to line 172. It is difficult to follow and interpret, but we want to support the effort. Many people realize that English is not his first language, and they see his struggling here as valuable mainly in helping him to develop fluency. This has led to a digression on what it means to teach science for nonnative speakers, a topic the group will probably bring up again after reading the teacher's case study.

Managing Criticisms

The impulse to criticize is especially strong when people see things as not going well; it's tempting to blame the teacher, even more so when the teacher has tried something that (most groups feel) hasn't worked: the students are still accepting the inconsistent answers. So conversations about the snippet to this point often turn in the direction of what the teacher is doing wrong, and, as we've discussed, this poses a central challenge for facilitating productive conversations.

The point is not to forbid questioning the teacher. To the contrary, it is important for people to talk about instructional possibilities, and that must include talking about what the teacher did. The point is to keep the questioning perceptive, thoughtful, and respectful, as well as, for our purposes, based on interpretations of the children's thinking.

We mentioned a criticism that comes up all the time in this segment: "She must not have demonstrated the experiment clearly enough." And we talked about how that carries with it an interpretation of student thinking we can examine explicitly: Does the evidence support the diagnoses that Ebony did not understand the question? Others might be that the teacher "missed the opportunities" of Rachel's and Diamond's comments (lines 25 and 64, respectively). Those, too, come from interpretations of student thinking.

So we want to talk about these ideas, but we also want to avoid naïve judgments of the teacher that fault her for not having addressed everything people find. Different observers will always notice different things; there were things the teacher noticed and considered that seminar participants miss. And teachers are always aware of more things than they could possibly address. Teaching means making choices, and making them quickly. Later, reading the case study, the group can find out more about Jamie's interpretations, both at the time of the class and looking back on it later; she, too, adds new ideas for what she might have done.

There may also be criticisms that do not come from interpretations of the substance of children's thinking. For example, someone might feel the teacher was paying too much attention to some students, Ebony in particular, and neglecting others. These can't become conversations about the substance of children's thinking, but it might be helpful to give them a little time. We try to recognize and respect the concern ("Your sense is that some children are excluded from the conversation.") but ask, "Can you imagine possible reasons she might have done that?" At some point, though, we try to bring the conversation back on topic.

The Snippet from Line 161 to Line 227

In almost all seminars we skip this segment, in order to get to what comes later. It is mostly made up of the teacher trying the experiment herself in front of the children. At the end there's some joking about how "maybe the gravity's tired" (lines 212–25), which could be interesting to think about. What's the joke? What understanding goes into making it or finding it amusing?

The Snippet from Line 228 to Line 288

This segment is DVD Scene 4. If time is tight, we sometimes skip this one too. It's only a couple of minutes long, though. Following are some points to notice:

- Starting with Jorge, Brianna, and Diamond (lines 253–57), the children said the book and the crumpled paper would hit at the same time.
- Diamond said that the crumpled paper was "like a ball" (lines 275–77) in answer to why she thought the paper would now fall at the same time as the book, which some notice as an example of her drawing on her experience.
- Allison said that the crumpled paper now "has a little more weight" (lines 278–80), the first mention of weight (or strength) since the day before. Some people may recognize an idea from Piaget's research in cognitive development, the preoperational reasoning that crumpling paper makes it heavier.

In a small number of groups there have been people familiar with the misconceptions research literature, including, specifically, the misconception that heavier things fall faster.* They've found this segment interesting, because the children's prediction is surprising: the research suggests most should think the book will fall faster because it is heavier. This has made for some interesting conversations: Maybe Jorge had seen this phenomenon? Maybe other children just believed him?

Others focus on the explanation that the reason the book and crumpled paper will fall at the same time is that the paper will be heavier. That sounds like reasoning in line again with the misconception. Did the children believe the crumpled paper would be as heavy as the book?

The Snippet to Line 367

We usually fast-forward the snippet until the children are all seated again. When we've skipped the previous segment, we give a quick summary of the bullets listed earlier.

This segment includes both the same sort of reasoning from the segment before, back to the idea of "strength" and "weight" affecting the speed of falling, and the wonderful new argument from Brianna that crumpling the paper doesn't change the weight. Every group that's viewed this snippet has noticed and commented on Brianna's reasoning, without any prompting, and it's typically the first topic of conversation.

Brianna's Reasoning

One likely reason people notice what Brianna said and did (lines 326–40) is that she was speaking to a misconception they've seen and thought about addressing, that the crumpled paper weighed more. Many will have had ideas for the teacher, that she should get a sensitive scale and weigh flat and crumpled paper, for example; here they see a child crumpling and uncrumpling the paper to argue that "it's the same size" (line 326). This can lead to a conversation about children's abilities and about teaching: that the teacher didn't address the wrong idea gave the children the chance to do that for themselves.

Another reason people notice this moment is that it's a textbook example of a "concrete operations" reasoning ability, "reversibility," in Piaget's stage-based theory of development. Brianna was literally reversing the effects of crumpling, as an argument that the amount of paper was conserved. If the topic had come up earlier, with the interpretation that the children were pre-operational, this would be new evidence that they were further along. In groups that have focused heavily on stage-based arguments, we might talk

---

* Heavier things do fall faster, but only when air resistance (as physicists would say) or "air blowing up" (as the children said on day one) is an important factor. If you drop two objects, such as a penny and a brick, on which air has little effect, you'll see that they fall at the same rate. At some science museums too, there are demonstrations of feathers falling at the same rate as rocks when they're in a vacuum. See the chapter notes, page 179, for a little more explanation.

here about how developmental psychology has been moving away from stage-based models.

The Same Strength

This time the children all agreed on the outcome: the book and paper fell at the same time. Brianna was the one to bring up the idea of book and paper now having the "same strength" as the reason they fell at the same time (lines 300–1), and Rachel explained that the paper had more strength because it was crumpled (lines 316–20).

People usually see these ideas as a return to mechanistic reasoning about strength or weight affecting how quickly the objects fell (as in the previous segment, if they've seen it, lines 284–86). If the comment doesn't come up on its own, we prompt for it: "How does the children's reasoning here compare with what we've seen them doing earlier?" This can lead to an interesting conversation about children's abilities—why is it that they're again showing these beginnings of scientific inquiry? Why had they stopped?

More than One Sense of Heaviness or Strength?

Until now, it had been reasonable to assume that the children were all thinking about *weight* and *strength* and *force* as meaning the same thing, but this segment may give reasons to question that; several groups have spent time on this.

Brianna had said that the paper and book had the same strength, but then she uncrumpled the paper to show that it was the same size—it was not heavier. One interpretation of this is that she changed her mind. Another is that when she was thinking about strength, she wasn't thinking of weight; she was thinking of some other sort of thing.

What did Allison mean when she said that her father could throw the crumpled paper to the ceiling but he wouldn't say it was light (lines 341–42)? Did she misspeak? Another interpretation some might consider is that she was thinking *light* could mean something other than how easy it is to throw something to the ceiling.

And finally, Diamond (lines 355–60) was trying to explain something about what balling up the paper did. It is hard to hear the word she used; did she say the crumpled paper had more "pages"? Might she have been trying to say something about the paper being more closely packed?

It's worth spending time on these questions, although with so little data, it's hard to come to any conclusions. Recognizing the questions is wonderful in itself and can raise new considerations for instruction. People might think of the following possibilities for instruction: spending more time on these ideas to help children articulate what they mean, asking students to come up with a definition for *weight*, and proposing a definition to them ("what a scale would read") to help clarify Brianna's claim.

The Rest of the Discussion

We've never made it to the last segment of the discussion in our workshops. Were we to get there, we'd expect to spend some time talking about the shift in the children's attention from "strength" to "shape" (line 376, etc.). Did the children resolve the issue of whether the paper was heavier? Did they have a sense of why shape would make a difference, or were they simply describing the change between the flat and crumpled sheet?

We'd also expect to spend some time thinking about Diamond's reasoning, as evident in her several comments about the paper being "balled up" (lines 373–74 and 405) and quick mention of how it could "roll" (line 387). Was she thinking of rolling as a mechanism for how a ball falls more quickly, or was she just using that to help explain that the paper was now round?

Jamie's Case Study

We've always watched the video before reading the case study, so these notes are based on that assumption. And we begin the conversation as usual with an open-ended question, such as "Who'd like to start?"

Several topics typically come up, including the choice of task and its connection to the county framework, particular aspects of Jamie's strategies, teaching science for ESOL* students, and Jamie's interpretations and responses to the children's reasoning. Our main interest, of course, is that last topic.

Interpretations of and Responses to Student Thinking

Usually someone raises the topic of how Jamie interpreted and responded to the children's thinking, either around a particular point or in general. If not, then at some point we do: "What sorts of things did Jamie write about seeing in the children's reasoning?" In essence, this is what we are trying to achieve with these materials—that readers will take up the practices Jamie is modeling in her writing of attending closely to the substance of children's reasoning. Along the way, we expect and hope that participants will take issue with some of Jamie's interpretations; to be sure, she had new ideas herself, looking back on the class. It is always important, however, to keep the conversation perceptive and respectful.

People generally talk about Jamie's reactions to the surprises of what Ebony and Rachel had to say about their results and of how others decided to accept the different outcomes. In the moment, she chose to focus first on the latter; she wanted the students to understand that in science, the results should be consistent. She also described the new interpretation she had looking back on the moment, after conversations with colleagues, about the possibility that

* English for speakers of other languages.

Ebony was playing with the question. Often that idea doesn't come up early in our seminars, and it makes a good topic here: What do people think about that possibility?

This interpretation made Jaime wonder about an earlier decision: if all the students knew so well what would happen if they dropped the book and the paper, might it have made sense to skip that experiment? Here is another case in point about a topic that might seem clearer in the abstract: most people think children should ultimately test all their conclusions with an experiment.

But in this instance, one argument goes, if any of us grown-ups were to conduct a fair test of the book and a flat sheet of paper, and the paper hit first, we wouldn't believe our eyes! We'd try to find the string that yanked the paper down so quickly, or the magnet, or *something*, because like the children, we already know what will happen. So the test might not be authentic, and we don't want children to think of science as inauthentic. On the other hand, it is important that children learn practices of conducting fair tests of ideas, and once in a while a test won't come out as they expect.

Another interpretation Jamie considered that could generate conversation is that maybe Rachel thought of gravity as variable like other things in nature, such as the wind. Is that plausible? And if that is what she was thinking, how might someone assess and respond to that idea? Would it reflect good thinking about gravity?

Teaching Strategies and Topic Selection

Very often the first things people notice are particular aspects of Jamie's strategies. One is the way she showed the children a video of their own discussion, an idea that several people have found intriguing, and not only for the purpose it served here; maybe it could help children become more aware of how they participate?

One group was particularly interested in how Jamie arrived at her topic, because it resonated with a couple of participants' experiences with thinking about science standards in their schools, which they saw as too low. At the other end of the spectrum, some high school teachers and college professors have seen the topic as too advanced for young children, based on their experiences. In one session, a professor was troubled by the possibility that the children would have the misconception that weight affects how quickly things fall reinforced.

Our main agenda in having the conversation is to promote practices of attending and responding to student thinking, so our first strategy whenever possible is to use the data of the case as a basis for discussions about what we might expect of children. What do these data about these students have to say with respect to the question? Do they suggest these children were, as Jamie expected, capable of more than the standards she described? Do they suggest that these children would leave with a misconception reinforced?

We're also interested to expand what people think of as the objectives of early science. It often happens that conversations slip back into assumptions that the bottom-line purpose is to arrive at the correct understanding, here that the children arrive at the conclusion that without air resistance, all things fall at the same rate. At least we should recognize the different objectives and consider which ones we'd choose for these children.

Changing Plans

Finally, Jamie's thoughts about changing plans midclass has led some people to talk about not being able to do that, either because of time constraints (if they do not move forward, they cannot finish the lesson) or because of school policies (they need to stick to the plan). They can't choose to make these sorts of adjustments in response to particular children's reasoning. In some instances, such as when we've had independent experience in the workshop participants' schools, we've had reason to question whether the options really are so limited, but we don't want those teachers' practices and choices to become the focus of conversation.

At the same time, we don't want to let time constraints or policies serve as reasons to avoid the hard work of understanding children's reasoning. From our perspective, teachers should be prepared to use their judgment well, and to make effective decisions, even if current constraints limit their options for responding. There's always *some* room for interpretation, and it's good to be ready to take advantage when opportunities arise.

CHAPTER 6

Eighth Graders Discuss the Rock Cycle

The rock cycle is a topic from geology, but please don't think you need to be familiar with it to work with this case study. We've used it extensively in workshops and seminars, and only a few people had already heard the term.

Basically, the rock cycle is to rocks what the water cycle is to water. So, water is always cycling through different forms and places on earth—falling from the sky, collecting into bodies of water, freezing into icecaps, evaporating or sublimating back into the sky, and so on. Similarly, rock is always cycling through different forms and places—bursting as lava out of volcanoes, hardening into solid *igneous rock*, eroding and breaking up into sediment and then slowly hardening again into *sedimentary rock*, changing over millennia into *metamorphic rock*, melting again into magma, and so on. This all happens on much longer time scales than the water cycle, of course.

Jessica Phelan was one of the most experienced science teachers in the project. She wrote this case study in the project's first year about her experience with trying a new approach to the topic, having the students work much more independently than they'd done in previous years. With this approach, Jessica's overt role in the case study is minimal—she spoke only at one point, with a brief suggestion to the students, about two minutes into the conversation. After that, although she was often monitoring the group's progress, she did not say anything at all.

Jessica's case study begins on page 97. We suggest you read the introduction, but stop before reading "A Shaky Start," which begins her account of what you can watch for yourself.

 ## Suggestions for First Viewing

As always, the transcript "Chaos in the Corridor" is available for printing from the DVD-ROM. In this case, there's so much overlapping talk by the students that you'll find the transcript has a fair amount in it that the subtitles simply leave out—we just couldn't cover it all and have the subtitles be readable.

The video is about fifteen minutes long. Once in a while in longer seminars (one and a half hours) we have time to watch all of it. We typically pause for discussion in three or four places.

1. *When Jessica says, "Can I make a suggestion?" (line 66 on the transcript), a little more than two minutes from the start of the video.* We then invite people to talk about what they have seen so far in the students' thinking and perhaps to speculate on what Jessica's suggestion is going to be. Sometimes we've paused just before she says that, to ask "What do you think about how they're doing so far?"

2. *Just after Jessica's suggestion (lines 72–73).* Whether it makes sense or not to pause here depends on the ideas after the first pause. If the suggestion is different from what the group has discussed, it makes sense to pause to think about how Jessica interpreted the students' reasoning.

3. *After Bethany reads her summary of what they have so far (lines 178–90), about seven minutes into the video.* That takes place almost five minutes after the teacher's suggestion, and it is a useful place to reflect on the students' progress.

4. *After Johanna's explanation with her shoe (lines 303–10), about eleven minutes from the start of the video.*

Be sure to read Jessica's case study after you've had a chance to watch and talk about the video. For commentary and a conversation guide, see the facilitators' notes on page 105.

■ ■ ■

Chaos in the Corridor

Jessica Phelan, Montgomery County Public Schools, Maryland

Every school year, I look forward to the eighth-grade geology unit, at the same time knowing that some of my students just aren't going to be into studying rocks. While some students find it fascinating, others just don't see the point in the topic. After all, rocks are just *rocks*, and what could be more boring?

This year, after several grueling periods observing and identifying igneous, sedimentary, and metamorphic samples, I felt that I needed to change something to get my students more involved. I decided to have them make a model of the rock cycle, and I decided to have them do it on their own.

The rock cycle, how rocks change over time from one form to another and back, is a standard part of our curriculum. I've had students make models of it before, but I've always allowed them use their books. Most would just copy a diagram onto a piece of construction paper, without really thinking about how one rock could become another. They didn't have to think logically, come up with their own connections, talk to one another, or use any creativity. Maybe I didn't trust that they would be able to come up with the cycle on their own. I wanted their models to be "correct," and I didn't want them to spend a lot of time on something they might get wrong.

This year, I decided that whether or not their models were correct was not the real point. I wanted them to have experience with making logical connections and devising a model for themselves, and to get some practice working on a team. As for our curriculum, all I really needed was for the students to come up with any cyclical process of rock transformations, and I was confident they could do that. They already had many components; they just needed to put them together.

Creating a rock cycle model in this way would take a lot longer than creating one using the book, but it turned out to be well worth the time and effort. For this case study, I'm going to focus on one team of eleven students who got off to a shaky start but went on to do better work than I had expected and, as I'll explain, better than I thought they were doing at the time.

Setting Up

A couple of days before the rock cycle activity, I told the kids to bring in "junk" from home that they could use to build a model, but I didn't tell them of what. I wanted to build a little suspense, and it seemed to work, since there was lots of talk about what the project was going to be. I also made it a contest, dividing the class into two teams, and told the students that the team with the best model would win a prize. Usually I don't resort to bribery, but this really got them excited.

On the first day of the model-building activity, I started class with two warm-up questions on the overhead: What is a model? and What are models used for? After we shared responses, I showed the students a small part of their last science talk on videotape, which had focused on the question How are rocks formed? That discussion had taken place at the very beginning of our geology unit, before any rock observations, and I wanted them to see how their ideas had changed. After playing the tape, I revealed that they would be constructing models of the rock cycle.

Until this point, their only experience was with the three types of rocks individually. They had observed igneous, sedimentary, and metamorphic rocks, and they had studied how each type formed. With this assignment, I was telling them for the first time that the three types of rocks could be connected in a cycle. (We had studied the water cycle earlier in the year, so I knew they had a general idea of what *cycle* meant.)

I told them they would be working in two teams and let them choose how to be divided up. They decided on one side of the room versus the other. Before beginning the competition, I told them that they would be judged on creativity and originality, group dynamics, and having individual rock cycle components. They learned the separate rock cycle components when studying the three types of rocks individually. (For example, igneous rocks come from cooled magma or lava.) They would need to tie these individual components together to complete the cycle. They were not allowed to use their books, but they were allowed to use their notes. I wanted them to plan the model using what they already knew about the three types of rocks.

They would have two eighty-minute blocks for planning and construction. I suggested that they appoint a leader and a recorder and that they spend the first block discussing the rock cycle and planning, and the second block building. It was very important to me that they talked through the cycle as a group before building their models, so that each student in the group would have a clear understanding of what the group was constructing. To me, working through the cycle, and coming up with connections between different types of rocks, was more important than the appearance of the final model. I told them that they should not use their building materials on the first day, because if they started using their materials, they might not focus on the cycle. On the third day, they were to present their models in front of the class.

I sent one group into a private corridor outside my classroom to work, and the other group stayed in the classroom. (The groups wanted to be completely separate from one another because they did not want any copying.)

A Shaky Start

The group in the corridor decided to start their talk by referring back to worksheets about the three types of rocks. Some of these worksheets were old homework assignments, and others were rock observation sheets. I think that they may have felt that they didn't know enough about the rocks on their own. So, at the very beginning of their conversation, Ryan asked, "Did anybody bring their sheets?" This sent the rest of the students rustling through their binders and then calling out pieces of information as they found them.

For most of this, I was in the classroom with the other team; it wasn't until I watched the videotape that I heard Ryan's request. When I did step into the corridor, it was just as Bethany was summarizing the discussion thus far: "So the Teutonic plates move and create rock, and then I have the igneous rock forms. Is that wrong?"

That didn't make sense at all, neither to me nor, I was sure, to them. We had not yet studied *tec*tonic plates, so I could only assume she or someone saw (and misread) the term on one of the worksheets. What followed wasn't any better. Lisa answered Bethany, "No, there's something. There was like de-desa- whatever," and Johanna helped her remember, "Deposition?" Neither of them seemed to feel comfortable with the word *deposition*, but they used it anyway, or tried to.

Hearing this bit of conversation made me think the students were not making any sense out of the problem. (Later, watching the tape, I could see that this kind of thing had been going on since they'd started.) They were using lots of big words, without understanding their meanings. This wasn't what I'd had in mind with the assignment! And it wasn't close to what these students were capable of doing. I had seen them use reasoning skills to solve problems before, and I knew that they understood the three types of rock better than they were showing. I felt that they had everything they needed to solve the problem of the rock cycle in their heads. It was just a matter of getting them to really think.

I was also struck by their lack of order and structure, with everyone talking at once. That was in sharp contrast to what I had just seen in the classroom, where the other team was off to a quick and efficient start, having elected a leader and started to attack the problem in a systematic way.

The Turning Point

As much as I wanted to let the students work through the problem on their own, I also wanted them to be successful. I decided to step in and say something (see Scene 2).

> **Teacher:** Can I make a suggestion?
>
> **Bethany:** Yeah.
>
> **Teacher:** You're looking at a lot of papers and using a lot of words that you don't know what they mean.
>
> **Gustavo:** Sure we do. [*Ryan laughs.*]
>
> **Teacher:** And if you're doing that, for your model, it's not going to be very good. So I want to start with what you know, not with what the paper says.

There was a pause, and then slowly they came back to life.

> **Johanna:** Well then we don't know anything.
>
> **Lisa:** Well the lava comes out. No, the lava comes out and it hardens—
>
> **Ben:** Yeah, the lava comes out.
>
> **Johanna:** So, so a volcano erupts.
>
> **Ryan:** Blam!
>
> **Lisa:** Then the, the lava shoots up.
>
> **Bethany:** OK. [*Pages ruffle as she discards the group's old notes and starts a fresh page.*] So lava—so a volcano erupts.

That was all it took, this one interjection from me, and the whole focus of their conversation changed. For just a moment they seemed reluctant, but they soon began talking about the rock cycle as a story that made sense, beginning with the eruption of a volcano. Starting over with a new set of notes, Bethany showed she understood that the first portion of their conversation didn't make any sense.

Johanna: Mm-hm.

Bethany: And lava comes out. Right?

Lisa: Do we have to talk about minerals too?

Johanna: No.

Ben and Gustavo: No.

Ryan: Nah.

Bethany: OK, so volcano erupts and lava comes out, the lava cools, and creates a, what?

Gustavo: A rock?

Johanna: An igneous rock?

Bethany: An igneous rock.

They were now off to a great start, having already identified one important part of the rock cycle, the formation of igneous rock. Here and later, it's interesting to notice that they were still using big words, such as *igneous*, but now they were using those words in ways that made sense to them.

During the class, I heard only about half of the corridor group's conversation as it was taking place, because I was traveling back and forth between the corridor group and the classroom group. I heard enough to know that they had taken my suggestion, but they still seemed to be terribly disorganized, talking over each other, and I wasn't expecting them to be very successful.

They were not polite, they did not take turns, and they repeated the same ideas over and over again. I was worried their model would show a lack of organization and planning. They were so much different from the classroom group, who sat around a table, took turns speaking, and voted on each decision.

Only after reviewing the videotape several times and discussing the transcript with colleagues did I appreciate what had taken place in the corridor. From the bits and pieces I heard at the time, I did not realize how much they really were working together, questioning each other and building on one another's ideas. Working out the details of their model, they raised questions of whether water was needed to cement sediments and whether a rock could be part sedimentary and part metamorphic. They produced and questioned the explanation that heat from the earth's core turns sedimentary rock into metamorphic (Ryan: "If there's so many layers, it's only gonna heat the first of 'em"). They questioned how rock could end up closer to the earth's core when more layers were piled on top—why would that move the bottom layer down?—and arrived at another idea instead, that the rock falls in when plates shift.

Eventually, they came to a point at which they realized they must reconnect metamorphic and igneous rock to complete the cycle. Arriving back at the beginning and starting the process over again was an important piece of the puzzle that they had not yet connected, and which they would not be able to find on any of their worksheets. How they worked that out is a nice example of the way they were working, talking over each other but thinking together.

Completing the Cycle

The corridor group's model so far was a linear progression from igneous to sedimentary to metamorphic.

Bethany: OK, now where do we go from here?

Gustavo: Igneous rock.

Ben: Right.

Ryan: We have to come back to the—

Bethany: Oh, wow—we have to go back to the—

Ben: Volcanoes, just a volcano erupts again because they—they don't stop erupting. They keep—

Lisa: Yeah, the same one erupts.

Johanna: That's why it's called a cycle.

Bethany: It's a cycle so it —

Ben: So it has to start over.

Bethany: The same rock has to start over again!

The beginning was igneous rock, so they had to get back to igneous, a small epiphany for the group. But how to complete the cycle? From this point in the conversation until the end, the students tried to figure out how the metamorphic rock could become an igneous rock, still thinking of a linear progression, but now trying to close the circle.

Ryan immediately had the idea that "the layers are gonna keep melting," but others were talking at the same time, and he couldn't get them to listen. Johanna came up with two different ways to explain how metamorphic and igneous rock could be related, both asking whether metamorphic rock could change to igneous rock on the earth's surface. "Does the metamorphic rock build the volcano?" she asked first, but Ben answered, "No."

A moment later Johanna came up with the idea of rock melting: "What if metamorphic rock is on, like, this ground, and then the lava goes over it?"

Ryan replied, "That's what I was saying . . . the lava will melt the meta-morphic rock."

But Keren and Bethany dismissed that idea; Keren said, "It'd still be a metamorphic rock." Did they think the lava wouldn't be able to melt the rock, or did they think the rock would still be metamorphic even if it did melt?

Eventually, Ben and Johanna came up with a great realization.

Ben: Well metamorphic rocks can be formed into sedimentary rock, but, but nothing can form into an igneous rock—

Johanna: Because igneous rocks—

Ben: —'cause it comes from out of the bottom.

Johanna: —comes from lava. W-w-wait. Magma—magma!

Johanna and Ben together realized that an igneous rock could only come from magma *underground*, "out of the bottom," and in order for the rock cycle to be a cycle, something must turn into magma. Until this point in the conversation, the group (with the possible exception of Ryan) had all been thinking of lava on the earth's surface rather than magma beneath the surface.

Johanna: Hold on, hold on. OK, you know how igneous rocks, like half—half of them can form from magma underground. And, and metamorphic rocks form underground. Could that have a connection? Hopefully?

Ben: Y-yeah, Ryan said that—that it'll, um, it gets closer to the um . . .

So maybe Ben had heard Ryan earlier, talking about how "the layers are gonna keep melting," but just didn't understand at the time. Now the entire group turned to Ryan and asked him to repeat what he'd said earlier, but by this time he was confused: "I'm lost. I don't know what you're [talking about]." While the group was prompting Ryan to remember, Gustavo seemed to be figuring out the process in his own mind.

Gustavo: What if, like, a sedimentary or, like, uh, uh, what, uh—

Ben: Metamorphic rock—

Gustavo: —sedimentary or a metamorphic melts and, like, and it, like, turn into lava, whatever, and then it dries up again.

Ben: —and the lava comes—

Gustavo: Would it be igneous?

Ben: —and the lava comes out of a volcano. And then it'd be igneous.

Gustavo figured out that if the rock melted and "dried up again"—he meant hardened—that would make it igneous. Ben agreed, adding that "the lava comes out of the volcano."

Lisa: Oh that's confusing.

Phillip: The magma comes out the volcano!

Gustavo: It's just one big cycle!

Ben: It's not confusing . . . it just gets more layers so it gets hotter, and, and it melts, and then it just—it comes out, back out of the volcano, and that's it. And that's it!

Bethany: Oh! [*Gasps*]

Johanna: Oh, I get it!

Bethany: Oh, me too! [*Laughs*]

Gustavo: It's a circle of life! (see line 472)

There was much excitement, as the whole cycle clicked in their minds. One at a time, my students realized that metamorphic rock melts into magma beneath the earth's surface, magma erupts from a volcano, cools, and forms igneous rock. For the rest of the period, the students began to discuss how they would build their model out of the things they had brought from home. They had figured out a cycle together, and they were ready to move on.

Reflections

I could have made the decision to interfere more in the corridor group's conversation. I could have told the students to take turns speaking or to vote on their ideas, as the other group did. I could have made suggestions or posed questions to lead them toward the "correct" answer. If I had done these things, how would their conversation have turned out? What would they have learned? Would they have had a better understanding of the rock cycle?

I'm still not sure exactly why I didn't do any more than I did. Maybe I wanted to be fair to both groups. I didn't want to help one group more than the other. Maybe it was because I saw that the students were on task even if they weren't organized. It took a lot of restraint on my part not to interfere more, but I feel my decision to let them work through the problem on their own turned out to be the right one.

In the end, I was pleasantly surprised when the corridor group's physical model and presentation showed real teamwork, creativity, and understanding of the cyclical nature of the rock cycle. That group's model actually received the highest score of any group in all of my five classes. The students' conversation may have sounded like chaos to me, but they were truly engaged in thinking and learning the entire time. Real conversation doesn't happen in a regimented and orderly way, so why should classroom conversation always be so controlled? Looking back, I'd have to say that what a teacher views as the most productive learning environment and the environment that students actually learn best in are not always the same.

As the end of the school year draws near, I think about how much I am going to miss this class when they are gone. Will I ever have a group of kids like them again? From the very beginning of the year, they always showed so much enthusiasm when talking about science. Did my teaching have anything to do with this, or were they just naturally curious thinkers?

Before our very first science talk, I set up rules for the students. One of those rules was to be a good listener. I told students that they should really be listening to what others had to say, not just waiting to talk. I realize now that this is a skill that I have also learned from them. As a teacher, I would often listen to students with the purpose of responding. I'd be waiting to talk, waiting for that perfect opportunity to step in and "teach," because that's what teachers do. Coming from a constructivist background, my idea of teaching has often consisted of questioning. The purpose of my questions was always to lead the conversation toward some logical conclusion that everyone could agree upon. Looking at this case study helps me realize how often I've had my questions in mind before I heard my students speak, so I never really had to listen to what they were saying.

■ ■ ■

Facilitators' Notes

Please see the general notes for facilitators in Chapter 3. Here we'll provide specific comments and suggestions with respect to talking about the snippet at the stopping points we suggested earlier.

Our purpose isn't to present a thorough analysis of the snippet but to give a sense of some possibilities for topics that might arise and topics a facilitator might choose to raise. So please do not feel limited to this list; other productive topics could come up as well. And conversely, please do not try to cover everything: No one of our seminars has addressed everything we describe here. It is better to have time to talk about some aspects of the video thoroughly than to rush through many topics quickly; there is value to lingering over ideas. There are one or two topics that we feel must come up, and we note them; all of the others we consider optional.

Two considerations for using this video in seminars: First, the participants will need a quick introduction, at least, to the question the students are trying to answer. The best strategy may be to summarize the information from the introduction to the case study. At minimum, explain that the students have studied igneous, sedimentary, and metamorphic rock, that they had earlier studied the water cycle, and that the assignment here was to come up with a model of the rock cycle.

Second, note that Jessica reflects on the final segment of the video in her case study. You might want to time things so that you get to that segment prior to assigning the case study for reading.

The Discussion to Line 66

Opening the Conversation

Stopping the video just when Jessica asks, "Can I make a suggestion?" makes for a natural entry into a conversation: You don't need to ask, "What will she suggest?" because everyone's already thinking about that. It's clear she's going to make a suggestion based on what she's heard, so it's easy to frame the question that way: "How do you think she's interpreted what they're doing?" or "What might she have seen in their work that made her decide to intervene?"

The other thing that helps start the conversation is the little malapropism of "Teutonic plates," which someone always notices. For people who know the term *tectonic*, this is a nice, obvious aspect of the data, and they enjoy pointing it out. (It's part of why we've also had an easy time getting the conversation started when we've paused the video just before Jessica speaks, when Ben says, "The deposit comes after that.")

It's tempting to play *Teutonic* for laughs (it means "German") as the leader of the conversation, but we recommend against that. It will get laughs anyway, which doesn't necessarily reflect disrespect, and in many groups it's clear it does not. But some groups sometimes forget to stay respectful of the students. Here as in general, it helps to just keep a neutral demeanor.

The other reason to stay neutral is that not all groups come to the same conclusion. Almost always, there's at least someone in the room who argues that the students are not making much sense, and typically others agree. Once in a while, though, we've seen a group decide that the conversation is going just fine, because the students are trying to use terminology or because they're referring to their source information. We see this as reflecting the participants' understanding of what science is, and part of our agenda is then to work toward a different understanding.

Most groups need prompting to come up with more specifics than "Teutonic." Once the conversation is going, it helps to start asking questions such as, "Can you point to something specific that makes you say that?" People generally don't have trouble finding instances that helped form their interpretations; the main reason to ask for specificity is to start establishing a norm of trying to back up interpretations with evidence in the data. It's easy to point to students' uses of terminology, for example. Some people point to "de- desa- whatever" (line 60) as evidence that the student was thinking about the next word but not the next physical mechanism. Some note the immediate move to the worksheets for information; some comment on Johanna's singsong manner in reciting "igneous rock forms, weathering occurs." And some have talked about how many times students referred to what they were "supposed to" do or think.

Another typical reaction at this point is for participants to talk about whether the students might have been distracted by the video camera. There is evidence to support that interpretation in the various comments about the cameraman* and the camera, as well as when students looked at and spoke to the camera.

The First Summary: "Teutonic Plates Move and Create Rock"
If no one brings it up, we make a point of directing the group's attention to Bethany's summary and asking participants to explain their sense of what it means: "So the Teutonic plates move and create rock, and then I have the igneous rock forms. Is that wrong?" The usual reaction has been that it doesn't mean anything; people see it as consistent with the interpretation that the students were speaking in terms they did not understand. Some people have tried to interpret a meaning. Either reaction is fine and useful for comparison later, when Bethany gives another summary.

Ideas for What the Teacher Might Do
One idea we've heard for how Jessica should respond (or prediction for what Jessica is about to do) has been to ask students to go back over their information sheets and review the ideas and terminology. Some people have suggested she could let them be for a while longer, on the view that they're starting to get settled and might put themselves on track. Some, too, have suggested various versions of telling the students to stop using the sheets and think about what makes sense.

As these ideas come up, it is important to consider how they may reflect subtle differences in diagnoses. One interpretation is that the students had not consolidated or fully understood the prerequisite background knowledge; that diagnosis motivates the action of recommending they review the information. Another interpretation is that they *had* the prerequisite understanding but weren't *using* it because they were looking to the information sheets instead; that motivates suggesting they stop using the sheets and think for themselves. Still another interpretation is that they were distracted and unsettled; left alone, they might find their way to more sensible ideas.

In one group we had an interesting conversation about the idea that the teacher should give the students a starting point for when the cycle begins, in response to the observation that the students were having difficulty knowing where to begin. Others objected to the idea, on the grounds that cycles don't have starting points, and the person who made the suggestion answered that the students would eventually come to see there was no starting point. Her interpretation was essentially that they needed a nudge to get going.

Jessica's Response (Lines 72–73)
Most groups will have little more to say here, and it isn't necessary to pause except perhaps to note that Jessica's interpretation and response were along the lines of something the group has discussed. She felt the students were taking the wrong approach, and she tried to help them by telling them to start from what they knew, as opposed to the information on the sheets.

* Seth Rosenberg, a visiting professor working with the project.

Some will have much more to say, though, especially if they have not considered this interpretation or response, and then it is useful to spend some time discussing her decision. We would not expect discussion here to reach a consensus or conclusion, only to air ideas. It's good to let this happen for a bit, but don't get stalled: when the video resumes, the students' discussion is much easier to follow for everyone.

The Discussion to Line 191

In most groups we've run, people have quickly seen the discussion as going very differently from earlier. They've noticed the shift in the way the students were sitting as well. Often we don't even need to ask an opening question, because someone says something in the vein of "Now they're making sense." The main thing to talk about, for this segment, is how the students now seem to be giving a straightforward, mechanistic account of transformations that start from hot lava cooling and becoming solid.

Again, we prompt for examples in the data to support interpretations. Here, the students' thinking is so clear and tangible that it feels repetitive to interpret what they meant by "the lava comes out and it hardens" (lines 76 and 78); "the rain and the wind cause, like, pieces—small pieces of the rock to break off" (lines 116 and 118); or the idea that the sediments would wash down and settle somewhere (line 140), then get "pressed together" (line 163) to make sedimentary rock. Bethany's summary at the end (lines 178–90) speaks quite well for itself, in contrast to what she'd read earlier about how underground "plates move and create rock."

Often, the participants reflect back on their earlier interpretations, whether to say, "Yes, we were right," or "That's not what we expected." If they don't, we raise the topic, not to embarrass anyone but simply to revisit the earlier thinking, as we would do with any prediction. It can be a useful time to talk about the uncertainties of diagnosing student difficulties— nobody can get it right all the time.

There may also be comments about the distribution of participation among the students—some are quiet and some are plainly not paying attention during portions of the activity. (There's background laughter among the boys, for example, during the exchange between Tracy and Bethany about sediment accumulating.) That's all stuff that matters in general for teachers, but it doesn't bear specifically on the substance of the students' thinking.

It has also come up that the students were still using the worksheets and the terminology—it can be interesting to talk about how they could use the same information in these different ways.

The Discussion to Line 310

Opening the Conversation
In the previous segment, the substance of the students' thinking was mostly straightforward and unproblematic. In this segment, there's more going on, and seminar conversations are much more varied.

Chapter 6

As always, we start by opening the floor to whatever anyone has to say, and it's typical for the first few comments to be generally evaluative: "They're still sense making," or "They're not very organized," or "I don't see much real thinking." We ask for specific instances or evidence to support these interpretations, as they come up. Does close examination of the video and transcript support the impressions people might have formed while watching the video clip?

Some, for example, feel the students were still using terminology they did not understand, such as *foliated* in line 242, and that they were still making rote use of their worksheets, such as in lines 274–75. Others have contested this interpretation, arguing that those lines are not representative of the clip as a whole. Some note evidence that the students were not always listening well to each other, in points that got lost along the way, such as Johanna's comment that "any rock" (lines 230 and 276) could undergo heat and pressure, not just sedimentary. One group focused on the brief exchange about whether the students had "gone over everything now" (lines 262–66), talking about that sort of decision as an important part of inquiry.

Contending with the Tendency to Evaluate the Teaching

It is easier with this case study than most to keep seminar conversations focused on student thinking, because Jessica is mostly (and literally) out of the picture. Even so, people typically want to talk about what she should have done. Some say the teacher (or the materials) should have provided more information about geological phenomena in preparation for this topic. Those with backgrounds in geology are more likely to go in this direction than others ("Haven't they studied subduction zones?"). Others say she should have given them a whiteboard or some other means on which to write their ideas so that everyone could see them; some think she should have assigned roles to the students or broken them into smaller groups, or stepped in to structure their interactions.

Our strategy here and in general is to ask that people articulate the interpretations that lie behind their suggestions: "What do you see happening here that makes you think that would help?" Most answers to that question are useful and push the conversation forward. Getting the interpretations out on the table allows them to be the topic of conversation.

One answer to the question might be that the students were confused and struggling to understand how the rock would get hot. We can then talk about that confusion, that is, focus on understanding what and how the students were thinking. Then we can return to the question of how to respond. We often find conflicting intuitions: On the one hand, seeing students confused makes people think that something's gone wrong, and they think about how the confusion could have been prevented. On the other hand, when the same people explicitly consider the question of whether it is necessarily a bad thing for students to struggle, everyone believes it is not. Talking about particular aspects of students' struggle in this case, and what might or might not be productive, brings the conversation back to our agenda to focus on students' thinking. Were they reasoning mechanistically? Were they working toward coherent understanding?

How Does the Rock Get Hot?

The main problem the students encountered was figuring out how the sedimentary rock got hot. Usually someone raises this topic, often by talking about the last thing that happened in the clip: Johanna demonstrated (lines 303–4) Keren's idea (line 295) with her hands and shoe.

We ask seminar groups to review the central issue; if they have trouble, we devote time to it and press for a clear summary: The students understood that sediments build up, one layer on top of another, which makes for a lot of pressure, but they didn't see where the heat came from. They had the idea that it was the heat from the earth's core, but they didn't know how any of the rock got close to that core: if the sediments pile up, that doesn't make the bottom go any lower.

It's useful, as always, to ask for evidence that this is what the students were doing, such as Ben's first mention of how the rock gets "lower and lower and then it gets hotter" (lines 196–97) and Bethany's question "So it sinks?" (line 201); Lisa's first posing of the question (lines 226 and 247); Ryan's mention of the earth's core (line 250); and Johanna's wry insistence to Ben that there was indeed an inconsistency in the explanation (283–86).

It's also useful to ask how the students solved it. Maybe someone will notice that Keren first had the idea, in line 295, which Johanna then illustrated with her hands and her shoe, in lines 303–10. (In one conversation, people had been concerned that some students were not paying attention, including Keren, who was mostly quiet. Naturally, seeing that she came up with this answer challenged that interpretation.)

What the Students *Weren't* Thinking

Some answers to the question "What do you see happening here?" concern what people see as missing. Of course, it's useful to be able to identify ideas students have yet to learn. It is important to recognize, though, that this is not assessing their reasoning as inquiry; it is assessing the completeness and correctness of their knowledge.

Because we are interested in helping educators assess science as inquiry, our concern is only that educators' sense of what the students *should be* thinking does not push aside the agenda to understand what they *are* thinking. Identifying what students are not thinking can help with that, because it rules out interpretations and may point out assumptions we take for granted that the students do not share. In the end, though, we want to bring the conversation back to understanding what the students are thinking. Often that needs a nudge: "OK, so you have a different sense from them about the problem they've encountered. What's *their* sense of the problem? Are they being reasonable, given what they know?"

In several seminars, talking about this segment of the video has raised another dilemma for us as facilitators: Some people have explained the puzzlement about how the rock gets hot as resulting from students' not under-standing that pressure causes heat. That can be helpful for interpreting the stu-dents' thinking: if someone assumed the students understood pressure to cause heat, it would be difficult to follow what they were trying to figure out.

The dilemma for us is that pressure does not, in fact, cause heat. Of course, it is a reasonable idea to suppose it does, and it is a plausible interpre-tation for what some students might think, that simply by virtue of being under a great deal of pressure, the rock would get hot. (Maybe that's why some of the students in the clip didn't see a problem?)

But should we digress to help the participants with their conceptual understanding of pressure and energy? It probably would not make sense to focus on that as a core objective, but it might be good to alert the participants to the issue. It's possible that this would give people added incentive to assess the students' thinking on its own terms—"Does it make sense?" rather than "Is it correct?" On the other hand, it could also make some people self-conscious or, worst of all, push them away from thinking of science as a refinement of everyday thinking, if their intuition is that pressure should cause heat and they aren't able to reconcile that contradiction.*

Tracy and Bethany's Exchange About Water (Lines 208–13)

The exchange between Tracy and Bethany often comes up on its own, just as part of a conversation over what the students were thinking. If it doesn't, depending on the time and the rest of the conversation, we sometimes raise it: "What were Tracy and Bethany talking about?" Here, too, the students were articulate and clear, so it is not hard to interpret their thinking: they were dis-agreeing over whether water was necessary to "cement" the layers of sedi-ment together. The rest of the students never took up the topic, and it might be interesting to talk about whether or not that's a lack a teacher might choose to address.

Ryan's and Lisa's Questions (Lines 292–93 and 311, Respectively)

The topic of whether a rock can be half-metamorphic continues into the next segment, and if there's going to be time to watch more of the tape, we don't bring it up here.

It often happens that someone else does, though, usually noticing Lisa's question of whether a rock can be half metamorphic. We generally facilitate by asking what Lisa meant by the question; what prompted her to ask it? Sometimes someone has also noticed Ryan's question in lines 292–93. He was asking how the entire rock, made up of many layers of sediment, would become metamorphic if only the bottom of it was under heat and pressure.

More often, though, seminar participants (like the students in the snippet) miss Ryan's question entirely; it can be useful to point it out, since one

---

* If the material under pressure gets smaller, then it gets hotter. That process of compression adds energy. So if you compress air into a smaller volume, the air gets hotter. But rock does not compress very much, even under great pressure.

reasonable interpretation is that Lisa was responding to his question: maybe the rock could be metamorphic on the bottom and still sedimentary on the top.

The Rest of the Snippet

We try to talk about the final segment, since Jessica focuses on it in her case study, in seminars that will read it.

It begins with the students discussing, just for a moment, the question Lisa repeated of whether a rock can be half metamorphic and half sedimentary (line 311), so it provides another opportunity to talk about this puzzle she and Ryan discovered. In some conversations we've lingered over Ryan's comment (line 318) that "if it's in the same area it's gotta be one thing," to consider the reason he might have had—if the rock is in the same area, it is exposed to the same conditions. Geologists have wanted to digress here, to talk about their answer to the question. They might bring up the question of size: how big were the students imagining this rock to be?

The students moved on from the question without resolving it, taking Ben's suggestion to "finish the rock cycle" (line 321), and this became the main topic for the remainder of the snippet: how does the cycle close?

There might be some conversation about the first interpretation of *cycle* as a repeated series, making "a volcano erupts again" (line 351) a plausible answer. The students quickly clarified, though, that "the same rock has to start over again" (line 357), and the problem became understanding how that happens.

Someone usually notices and mentions Ryan's answer, that the rock will melt again (line 358), and how he tried several times to get the others to pay attention (lines 366 and 382). Eventually, they came back to what he was saying, at Ben's prompting (line 437), and they settled on that as their answer.

Along the way, they considered and set aside a number of other possibilities, shopping for ideas, including the idea that the metamorphic rock builds a volcano (line 372), which they rejected because it still didn't mean the same rock starting again (lines 374–76), and the idea of lava flowing over rock on the ground (lines 380–81), both ideas from Johanna. One interesting comment to interpret is Ben's idea that "nothing can form into igneous rock, because it comes from out of the bottom" (lines 422–25). What do people think he meant by that?

In the end, many people want to assess the students' model. One line of reasoning is to compare it with what is known in geology, and if there are geologists in the room they might compare it with what they know. Another is to think about whether the model is sensible and complete given what the students knew. Some participants have noted, for example, that the students could have realized that igneous rock or sedimentary could melt back to magma, too, instead of taking the cycle as a single circle: molten rock to igneous to sedimentary to metamorphic and back to molten.

Jessica's Case Study

We have always shown and talked about the video before assigning the case study, for the reasons we've mentioned: it helps make the case more real, and it gives participants a chance to have and talk about their own impressions before reading the teacher's.

Conversations about Jessica's case study tend to have a different feel from conversations about other cases. There's more of an empathetic tone, with comments that add to or elaborate on Jessica's thinking, instead of the critical stance we've come to expect. That's probably at least in part because of the small role she plays in the snippet. As she explains, during the class she was splitting her time between watching the group in the classroom and watching these students in the hallway. One effect is that Jessica is writing from a similar position to seminar participants', finding out about the students' thinking from watching the videotape.

The different tone might also be because Jessica is so thorough and forthcoming in explaining why she chose to act as she did. In other cases, there are many moments of teacher actions that readers might want explained, and it's not possible to elaborate on them all. Here, there are only three places to talk about the teacher's choices, and by this point, every seminar has already talked about one of them ("start from what you know"). Generally, participants raise the other two, which Jessica discusses in the case study: her thinking about the assignment itself and her decision not to intervene further to do something about the "chaos" of the group's work.

The Assignment

Jessica writes that the "real point" for her in the assignment was to get students to think for themselves, making their own logical connections. All she needed was "any cyclical process of rock transformations," which she was confident the students could accomplish. Teachers recognize and discuss the theme of trying to balance independent thought and covering content in their assignments, with the difficulty always of deciding how much time to allot to each.

Inevitably, conversations turn at some point to state standards and the required ideas. Many of our seminars move to this quickly, with people wondering aloud whether they could afford to give this sort of assignment and have this sort of flexibility, which everyone seems to think would be a nice thing to do. Many state standards include sections on students' abilities for independent thought, including developing models, but the general sense of our conversations has been that thinking about standards has teachers doing less in the way of student-run explorations.

The conversation could easily stay on the topic of whether teachers can afford to devote such time to such open-ended tasks and so little coverage, given the constraints of standards and assessments, and never return to the case study. So we let the conversation wander for a bit but then pull it back.

Deciding Not to Intervene

Often people talk about what Jessica didn't do, namely, intervene to help the students become more organized and attentive to each other. Jessica expected this "chaos in the corridor" would not produce as good results as the more orderly group in the classroom—she talks about having to restrain herself—and that aspect of her account resonates with teachers' experience: it's hard to watch students and not try to fix things for them, but, people tend to say, it's important sometimes to hold back.

Jessica also talks about how this disorderly group produced the best model she saw from students in several classes. During class, she'd had the opposite sense, that these students were not working very well together. Afterward, she thought differently of that assessment, writing, "Real conversation doesn't happen in a regimented and orderly way, so why should classroom conversation always be so controlled?"

We raise that topic of conversation if nobody else does. Our core purpose in these case studies is to help educators think about what productive inquiry looks like, and Jessica's point here is provocative: none of us would want to discourage real conversation about science, but would any of us be less liable than Jessica to miss it? Looking back, with the evidence of what the students produced, she questions her judgment at the time that the students weren't doing well. That judgment, of course, was in line with traditional expectations, which don't really allow for uncontrolled discussions like the students' in this snippet. (Sometimes our seminar conversations become lively and spontaneous enough that we can see Jessica's point firsthand. Then we can pause to ask, "Would a conversation like the one we're having now be OK in school?")

The Teacher's Interpretation of the Problem

The topic of Jessica's interpretation of the problem seldom comes up unless we raise it, probably because everyone feels we've already talked enough about how Jessica responded to the students' initial approach. The main reason possibly to revisit it is to reflect on her advantage as the teacher. When she stepped into the hallway and heard "so the Teutonic plates move and create rock," she could draw on what she knew about the students from lots of other work to arrive at her sense that this discussion "wasn't close to what they were capable of doing."

We sometimes talk about that advantage and what it means for having snippet conversations with teachers about their students, for participants interested in doing this sort of thing themselves. When in doubt, it is a good idea to defer to the teacher on questions of interpretation, because she has much more knowledge of the students. That doesn't mean we should defer in general—we have some data before us, and we should use it as evidence. But that evidence is often open to more than one interpretation. In this instance, of course, we have the evidence of what happened as a result of Jessica's intervention to support her interpretation.

Back to the Discussion: The Students' Thinking About How to Close the Cycle

Another topic that usually doesn't come up unless we raise it is Jessica's account of the students' thinking about how to close the cycle. We don't always do that, because it feels a little like going backward to return to talking about particular student utterances again as we were doing earlier when talking about the video.

But there are reasons to consider raising the topic. One is that it can be useful to look back on the discussion and think about what made it productive, in light of Jessica's reflections about "real conversations." Another is that Jessica raises interpretations and questions that probably differ from seminar participants'. For example, looking back at Ben's statement that "nothing can form into an igneous rock because it comes from the bottom," Jessica credits him as having contributed to the insight Johanna voiced that magma came from underground, when they had been thinking only of lava on the surface. Finally, it can be nice to return to this level of attention to student thinking as a coda on the seminar that connects to Jessica's final thought. What does it mean to really to listen to what students are saying?

CHAPTER 7

Third Graders Discuss Bubbles

Pat Roy joined the project during the second summer, and this was her students' first time having an open discussion in science. Right away at the start of school year, her class was scheduled to work on the county's third-grade task "Bubble Trouble," and Pat decided to introduce this different kind of talk in which she would encourage the students to express and discuss their ideas.

Her case study begins on page 117. As she explains, she had assigned the children some writing about the question for homework: What will happen if you blow bubbles through differently shaped wands? If you read just the introduction, you'll stop before Pat tells you about what she'd seen in the students' answers. Read two more paragraphs if you'd like to find out about what they wrote before watching the video. After that, Pat starts describing what's on the video.

We've found this case takes a little more careful attention and patience than the previous ones for people to hear and have a sense of what the children are doing. When we've tried it for the first case study, we've had trouble getting a conversation going—people just don't have much to say. So we now generally hold off using it until they've had more experience listening to children's reasoning.

The transcript for "The Trouble with Bubbles" is available on the DVD, and the facilitators' notes start on page 126.

 Suggestions for First Viewing

The video on the DVD is about half an hour long. When we use it in seminars and workshops, we generally stop it in three places for discussion.

1. *After Pat reviews the question (lines 11–14).* Somehow everyone always knows that the bubbles will all come out round, whether from experience or from a sense of how bubbles work. It's useful to try to talk about the latter, because it isn't easy to do! Exactly why do they come out round? Mostly, though, we ask participants to talk about what they expect to hear from the third graders.

2. *After Lester says he tried it before (lines 89–90), about four and a half minutes into the video.* What did you notice about the students' reasoning?
3. *When Pat calls on Jerome again (line 198), about ten and a half minutes into the video.* We used a voiceover for "Jerome" (not his real name) because we did not have consent to include his face and voice in the video. (The voice belongs to Matty Lau, a graduate student research assistant, who was doing her best to mimic the boy's tone and pace.)

With experienced groups, that's been enough for an hour-long conversation. Of course, there's much more in the rest of the snippet, and if you have more time you should continue. One elementary science methods seminar spent two sessions on this case, which gave participants time to see more video, and the added time led to greater depth in the group's conversation and appreciation for what the children were doing.

■ ■ ■

The Trouble with Bubbles

Pat Roy, Prince George's County Public Schools, Maryland

This was our very first try at this new kind of talk, so we were very excited!

My class consisted of twenty-three third graders—eleven boys and twelve girls. About half of them started the year reading below grade level and fifteen of them qualified for free or reduced lunch. Four students were of Hispanic origin, two were white, five were biracial, and the rest were black. Our Prince George's County school is within blocks of the district line and is quite transient. (Three of the students in this talk relocated during that school year.)

It was the start of the year in 2001. Prince George's County required that each grade level complete PLUS (Performance-based Learning and Understanding in Science) Tasks each quarter. The initial purpose of these tasks was to create activities that would be similar to the MSPAP (Maryland State Performance Assessment Program) that the state administered to third, fifth, and eighth graders.* The county selected the theme "Bubble Trouble" for third grade during the first quarter. The activities had performance and writing components that were done both in groups and individually and recorded in booklets. Each task began with an engagement, which introduced the theme, then a prediction, which the students wrote in their booklets, and then the experiment.

Generally, for our science lessons, we always followed the task booklets, starting with the engagement. There would be a little bit of discussion then,

* The MSPAP has since been replaced by another exam.

but just for motivation. Then we'd move right into the prediction and then the experiment, without any more discussion. We never really talked about what the students were thinking. For our first science talk, I decided to have the children discuss their predictions, using the question in the booklet. That was the science lesson for the day; the experiment took place the day after.

This talk was going to be different from anything we had done before. I followed what other teachers in the project had done and told the students they did not need to raise their hands to talk. That intrigued them, although many of them raised their hands anyway. We established some ground rules, which basically consisted of everyone being respectful and considerate when others were speaking. We decided that only one person could talk at a time.

I was concerned with how this talk would take off, because I could not remember having a class with so many quiet and shy students as I did that year. I figured about half of my class would probably participate. This didn't give me a lot of confidence!

The Science Talk

The task began on September 4. We were to investigate whether differently shaped wands would produce differently shaped bubbles. I had made wands (a square, a triangle, and a rectangle) out of small straws, with regular circular wands as the fourth shape. I had also created a worksheet, separate from the standard task materials, on which I drew all of the wands we would be investigating. I showed the students the wands, and I asked them to write what they thought would happen if they tried to blow a bubble through each of them.

I expected most of the students would think that the shape of the wand would determine the shape of the bubble. I couldn't imagine any of them not having been exposed to blowing bubbles through circular wands, but I didn't expect many would have seen bubbles from wands of other shapes. From their written responses, going into the discussion, I found that two students didn't believe that the differently shaped wands would change the shape of the bubbles. Another student wrote that the bubble blown through the circular wand would come out as a circle, but if you tried to blow bubbles through

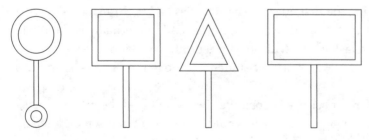

Bubble Wands.

the other shapes, the bubbles would pop. The rest of the students wrote that the shape of the wand would determine the shape of the bubble.

To start the talk, I held the differently shaped wands in my hand and restated the question: "What would happen if you blow a bubble through a rectangular shape, a triangular shape, square, or circle?"

Kedra said she thought "it will come out in the same shape" as the wand, and everyone started talking at once. I reminded them of our rules, and Louis was next.

> **Louis:** I think it will come out . . . I think it will come out in a circle always.
>
> **Teacher:** You think all of these will be blown, when you blow through it, the bubble will always be a circle?
>
> **Louis:** Yes.

I should have asked Louis why he thought they would all be circles. I was probably unconsciously thinking that since he had the correct answer, there was no need for an explanation. Louis was one of the students I had expected would realize that the shape of the wand would not affect the shape of the bubble. It didn't seem to bother him that he was the only one so far who predicted the bubbles would all come out round. He liked to debate or discover something that no one else had discovered. (At least he thought no one else had discovered it!) I don't want to portray the wrong image of him; he may have liked to be right and would debate with others, but he was a fun-loving student. He didn't put others down because they didn't seem to have it. I enjoyed him a lot!

Odessa commented that he disagreed with Louis; he thought that the wands would affect the shape of the bubbles. I was excited that he was willing to take a risk and share what he thought. It still seemed that he was lacking confidence, as I had to really pull out of him what he was trying to say. Even then, it wasn't audible enough to record his response on the tape. I told him that his classmates could not hear him, because I wanted them to get used to speaking to each other and not to me.

After Odessa, Jasmine agreed and asked how a bubble could come through as a circle if the wand was a triangle. I was really interested in hearing more of what she was thinking, but Zoë interjected her disagreement. I probably should have gone back to Jasmine's thought to see if she would have said more because she generally gave careful thought to what she wanted to say.

> **Zoë:** I disagree. I don't think that . . . the rectangle, the square, and the triangle will come out in the same shapes. . . . In that shape, because it, it can't get all the sides that it suppose to get. If it was blown in that, it would just be, it would be like it is flat.

Teacher: But, if I would take this triangular shape and dip it in the bubble solution, it's going to cover all of this, so why don't you think it'll blow out triangular? Or do you think it will?

Zoë: Maybe it will or maybe it won't.

Teacher: [*Laughs*]

Zoë: I don't know.

Did I confuse her? Zoë was the one who had written that the circular wand would blow round bubbles but that the bubbles would pop if you tried to blow them through the other shapes. It was not clear to me at this point what she was saying about not getting the sides it was supposed to get and that it would be flat. I was trying to clarify when I described to her that when the triangular wand was dipped in the solution, it would fill the wand, sides and all. When I watched her on the videotape, her response reminded me of what I sometimes do when I think I have a hold on an idea and then some other idea enters in, and then I am not sure what I think! I wonder if she thought that I was trying to get her to think that the bubble should come out in a triangular shape. I really should have asked her more about what she meant.

The conversation continued with Samuel.

Samuel: I disagree because I tried it out in a triangle—

Teacher: [*To other students*] Wait a minute. [*To Samuel*] Go ahead.

Samuel: I tried it with the triangle before and it came—

Teacher: You tried it before?

Samuel: And it came out in a circle.

Teacher: So you're going on past experience.

Louis: I've done, I've done it in past experience too.

I did not want these students' experiences to end the discussion, and I wanted them to start giving reasons for their predictions.

Teacher: You have done this before too. Hmm, well, suppose I don't believe that. Suppose I think that if I put this square shape in bubble solution, my bubbles will come out as a square; suppose I think that.

Then Lester said that he had tried it with a rectangle and the bubble came out a rectangle! But Samantha didn't believe that.

Samantha: If I tried it, um, and it did came out like, like the shape, like the shape that it was supposed to look like, I still wouldn't believe it.

Teacher: You wouldn't believe it? Why not?

Samantha: Because the corners, it's a corner, it'd be silly. Because then if you try to touch the corners, it'd pop.

Kedra: True that!

Teacher: So now, explain that for me. What do you mean by that?

Samantha: Well . . .

Teacher: Do you need to use one of these [*indicating the wands*]? [*Samantha shakes her head.*] No?

Samantha: If I tried it, and it did come out like it, it would be weird because it wouldn't be a square, or, it, I mean it wouldn't be like a corner shape or, or a point.

Samantha was a very vocal student who could be quite dramatic. Sometimes she seemed like a miniature adult in the way that she expressed herself. I wonder if she said it would be weird to have the bubbles come out in the shape of the wands because she had never seen bubbles in different shapes. The way she said she wouldn't believe it even if she saw it was an example of her adultlike choice of words. Even if she saw it, she would still question how that could happen! In her mind, it was impossible!

Then Zoë, who was also quite thoughtful with what she said, tried to clarify what Samantha had said. The same girl who wasn't sure what she thought minutes before but could now explain what someone else was trying to say!

Zoë: I think I understand what she's trying to say.

Teacher: What is she trying to say?

Zoë: I think she's trying to say that if it, it was in a square thing—were you talking about the square?

Samantha: I was talking about the whole thing.

Zoë: Oh, I think, I think I know what she's, I think I know what she's talking about. I think she means that it won't come out, in the, like a, a real, like the way a real square suppose to be.

Teacher: And can you tell me why you don't think so? How do you think it's going to come out?

Zoë: I think it might come out as a circle bubble.

Teacher: For all of these?

Louis: Uh-huh.

Jerome: When you blow the, the circle for a bubble out, it'll come just like a circle. And if you blow a rectangle out, it will come like a rectangle. But if, if it come out like a rectangle, it'll pop.

Jerome thought that the bubbles could be blown through the differently shaped wands and come out in that shape. But he thought they would pop because of the corners. It seemed that Jerome and Samantha were associating corners with sharpness. From their prior knowledge, they would know that if you touch a bubble with a sharp object, the bubble will pop.

Jasmine: If it comes out like a circle, how come it can't come out like a square or a rectangle or a triangle?

Teacher: That's a good question. That's what I'm trying to find out.

Jasmine: When you blow a bubble out of the circle, it comes right out just the same.

Teacher: OK, so what are you saying?

Jasmine: Then how come it can't come out with corners?

Teacher: I didn't say it could or it couldn't.

Zoë: I don't agree with Jasmine. I think she's wrong because, um, a circle doesn't have corners, so how, how would that, how would those shapes, the bubble come out in corners?

Teacher: But she was saying that if the bubble can come through the circle as a round bubble, then it should come through these the same shape as the wand. Is that what you're saying?

Jasmine: Yes.

Zoë: No.

Teacher: You don't agree with that?

I didn't know what Zoë was trying to say, but Samuel jumped in with something else.

Samuel: I don't agree.

Teacher: You don't agree?

Samuel: I think because, like, probably a oval would come out and an oval, because it doesn't have corners. Because those have corners and that's why it probably doesn't, it can't come out.

Now this was quite interesting. Samuel brought up the idea of an oval shape, which was not one of the shapes that we were investigating. I hadn't thought of the possibility of shapes other than the ones I had shown them entering into the conversation. Samuel was associating the oval shape with the circular wand, as they both are without corners. He thought an oval-shaped wand would produce an oval-shaped bubble just like the usual round-shaped wand produces a round bubble, but the shapes with corners would not work. Since he knew the shapes with corners wouldn't work, he came up with another shape similar to the circle that, in his mind, should work. It was not so much whether he thought the shape of the wand would or wouldn't change the shape of the bubble, but that the corners wouldn't even allow the bubbles to "come out." I hadn't anticipated that idea.

Teacher: So you, you think having corners will affect whether it will come out in the shape that it is?

Many Students: Yes.

Samuel: It probably won't come out because it has corners.

At this point, Odessa contributed to the discussion again, saying he believed that the bubbles would all come out in the form of a circle. It seemed that many of the students were beginning to change their previous thoughts about the bubbles coming out in the shape of the wands. There was some thought that the bubble would be blown through the differently shaped wands but they would pop because of the corners on the wands.

Teacher: OK, Rondell?

Rondell: I disagree because, um, when Brian said . . . that when the bubble comes out as, um, the square then it, it going to [be] the square, but it's not going to be the square because, um—

Teacher: So are you . . . saying that you don't agree that the bubble will come out the same shape as the wand? You disagree with that? What shape do you think it will come out? Do you think it will come out? Do you think this will, it will even work?

I wonder what would've happened if I'd stopped after my initial question. Sometimes, too many questions can interrupt the students' thinking, but I am not sure if that happened here or not.

Rondell: I think it will because it will just come out with a bubble.

Teacher: What shape of the bubble?

Rondell: A circle.

Then Kevietta got into the conversation and commented that she believed that the bubbles would come out in the shape of the wand, because she remembered that had happened on a TV show, *Barney*, that she saw when she was three. I prompted Rondell at this point since he had been speaking before Kevietta's response. I wanted to make sure that he'd had a chance to complete his thought, since he generally was not a participant of classroom discussions. He continued and expressed his disagreement with Kevietta.

Rondell: I disagree because, um, once it blows . . . if it's like you put it in . . . a thing, and if you blow it, it's going to come out like the same thing, because it's, um, . . . it's going to change into a bubble.

This was another very interesting idea. Rondell was quite thoughtful here. I thought that he and other students were saying that the bubble would come out as the shape of the wand and then transform into a round bubble, so I decided to ask them about that. I hoped this would get them talking in more detail.

Teacher: So you think when you blow through it, it's going to first come out the shape of the wand, and then it's gonna turn into a bubble? At what point will it turn into a bubble, a round bubble? At what point do you think that's going to happen? When, when will that happen? Will I be able to see it go from a square to a circle?

Many: No.

Teacher: When do you think it's going to happen?

Rondell: Um, like, if it, like, blow, if, if you blow it and if it's the, um, same thing as the square and if it, if it . . . changes into a circle, it's going to pop.

Teacher: Say that again, I'm sorry.

Rondell: Like, if you, um, blow the, um, square . . . into the same, um, thing it is, then, um, when you, um, when you blow it, it's going to change into a bubble and then pop.

Teacher: Right, but when will it change?

Rondell: Just at the—

Teacher: Will you be able to notice when it changes?

Rondell: No.

Teacher: Why not?

Rondell: Because if it change out as a, um, as a square and if you don't see the square change out, and if you blow it . . . it's just going to be like a bubble.

Teacher: A round bubble? So it won't come through this other side as a square?

[Rondell shook his head "no."]

Teacher: At all? So at some point, you're . . . saying that at some point it's gonna be square and then as it comes out on this side, it's going to be in a circle? OK.

I believe Rondell was saying that when the bubble solution was just sitting in the wand, it would have the form of the wand; however, when it was blown, it would blow out as a circle and pop more or less immediately. This was interesting because it seemed similar to his initial idea but different from his response to Kevietta. Was he changing his mind back and forth or maybe holding several ideas at once somehow? These are the kinds of questions I was asking myself as I led this science talk. So much was going on that it was difficult to pay attention to everything, but I certainly saw at the time that there were many great ideas coming out and the kids were doing plenty of great things to respond to each other's ideas.

The discussion continued for another ten minutes, and the next day I gave the children the opportunity to experiment with the differently shaped wands and the bubble solution. My greatest regret in this science talk is that I did not videotape the actual experiment of blowing the bubbles through the different wands! I hadn't done much videotaping of class sessions, so I wasn't quick in my thinking of capturing everything on tape. Now I know!

New Ideas

I learned things about some of my students through this science talk that I may not have noticed in a regular science class. For instance, I didn't know Rondell and Odessa could participate so well. Since I was not pressed for time, I could be flexible and let the conversation go as long as the students needed it to and were on task. Plus, I was able to be much more observant of their thinking than in other lessons. Did they sense the relaxed "this is the students' talk" atmosphere more than in our regular lessons and feel less inhibited? Did my role as a facilitator affect the way in which the students viewed this lesson?

As I looked over the students' written responses prior to the discussion, I separated the papers that predicted the bubbles would come out in the shape of the wand and the ones that said the bubbles would come out round. I also categorized the ones that mentioned trying it before. Rondell had written that the circle, square, triangle, and rectangle were all going to make a bubble. The way he wrote it made me think that each bubble would take on the shape of the wand.

A year later, after watching the video again and looking at Rondell's paper, I realized that he hadn't changed his prediction; I had misinterpreted his paper. In the discussion, he said that the bubbles would come out as a "bubble." The more I listened to him, the more I realized that when he used the word *bubble*, he meant round. When I asked him to clarify what a bubble meant, he said round. So I went back to his paper and reread his response, and sure enough, his prediction seemed to mean that the bubbles would all come out in a circular shape.

Last Thoughts

I was very critical of myself after this science talk and really didn't like it. First of all, I didn't like seeing myself on video, and then I felt I was doing too much talking and not allowing the students to do their own questioning of each other. However, as I looked at the video more and began to focus on what the students were saying and not on myself, I began to value the lesson.

Considering the fact that this was our first science talk, it was appropriate to demonstrate and model for the students what I wanted to happen. I shouldn't have expected them to automatically know what to do, so it was OK that I did the amount of talking that I did. As I looked more closely at the students' interaction, I actually saw some great things happening.

The students were listening to each other and clarifying what others were saying. They were respectfully disagreeing with each other and supporting their thoughts with evidence and logic. Some of the students who participated were children who didn't express their opinions in other class discussions, and to hear them contribute without a prompt from me was so exciting! Considering that this was fairly early in the school year, they did a great job of thinking and sharing their thoughts with each other. Later on in the year, I used the science talk format to discuss classroom issues, which really seemed to make a difference in student accountability and overall respect for each other.

I've now made it part of my instruction to have a talk like this before introducing a new concept. The students like it, and it really prepares them for the lessons more than just doing the engagement activity from the task booklets provided. With the science talks, it seems that I have to do less explaining, because the students bring up so much in the discussions.

■ ■ ■

Facilitators' Notes

Please see the general notes for facilitators in Chapter 3. Our purpose here isn't to present a thorough analysis of the snippet but to give a sense of some possibilities for topics that might arise and topics a facilitator might choose to raise.

What's There to Say About Why the Bubbles Come Out Round?

We have not had extended conversations about why the bubbles always come out round. Our main purpose in visiting the topic is to help everyone appreciate the substance of the question beyond knowing the answer: science is about more than knowing what happens; it's about understanding the mechanisms that cause things to happen.

The group may think about what exactly a bubble is: Is it a *volume* of material (as in a round water droplet) or is it a *surface* of material (hollow like a basketball)? Some people talk about the soap film as the material, noting that it is not a solid; some want to digress into the question of whether the film is therefore a liquid. Some people are familiar with the term and concept of *surface tension*. That idea can connect to other phenomena such as insects and needles floating on water. Some have seen nonspherical bubbles, either sitting in wire frames to support their shape or because they are in multiples, and the question can shift to why those bubbles are not spherical.

It is not important whether or how the group comes to a resolution; all we need is that they talk about some plausible causal mechanisms. And it is a wonderful sign when a group sees so many directions to go and connections to other phenomena that we have to ask them to stop! But that's what we do after a few minutes, especially if time is limited, so that we can get to the video. (See the Notes, page 180, for a little more on bubbles.)

The Discussion to Line 90

Opening the Conversation
We pick line 90 as a stopping point because it is a provocative moment, with students supporting different answers based on claims of having tried to make bubbles with differently shaped wands. With less experienced seminar groups, the first reactions are usually about what the teacher should do, and so we ask for what perceptions and interpretations led to those ideas.

Thinking About Hands-On Opportunities
Another topic that often comes up at this moment is the concern that students need hands-on experiences as a prerequisite for conversation. Pat has announced they will experiment with bubbles, but "not today," and sometimes in our workshops someone questions the value of proceeding with the talk, especially with so many students saying the bubbles will come out the same shape as the wand (lines 15–16, 38–41, and 47–52).

Some of this is just a general sense among elementary educators that science has to start with experiments and science activities have to be hands-on. It might be good to take a moment to discuss that topic (we talk about it a bit in Chapter 10). Here, though, the concern for experimentation comes in part from what people have seen and interpreted in the students' reasoning.

In particular, several students claimed to have tried the experiment and found the answer. Samuel (lines 79–83) said he had tried a triangular wand and the bubble came out round; Louis (line 85) said that he'd tried it too and agreed. But Lester said he'd seen a bubble come out rectangular (lines 89–93). (Samuel then responded that he "didn't try a rectangle" (line 94), so he couldn't speak to Lester's results.) At this point, the discussion is about empirical findings, and people often see that and suggest a reasonable option: stop now and have the children try making the bubbles so that they can agree on what actually does happen.

Sometimes, though, participants suggest that's what the teacher should have done, and we don't let that conclusion go unchallenged. It may reflect a view either of children or of science that empirical tests are the only way to make progress. Perhaps the children had reasons for thinking what they did other than what they remembered having seen. If we think they might have had cause-and-effect reasons for their beliefs, then we might want to help them articulate those reasons.

Interpreting Zoë's Idea (Lines 69–70)

Sometimes someone identifies Zoë (lines 69–70) as having started in that direction; when no one does, we point it out: "What do you think of Zoë's idea, that 'it can't get all the sides it's supposed to get'? What do you think she meant?"

People have had different ideas about this. Some think she was trying to say that the bubble couldn't form all six sides it would need to come out as a rectangle. That she said "it can't get *all* the sides" may mean she was imagining the bubble could get *some* of the sides. It can be helpful to try to fill in that reasoning: if we imagine the bubble coming out a rectangle, maybe we can think of the lateral sides forming but not the ends. Others have focused more on the phrase "it would be like it is flat," thinking she was talking about the soap film while it was in the wand, having only one side, and maybe that's the only side it would get. It isn't possible to pin down precisely what she meant, but most see her as trying to think about how bubbles form.

Filling in the Mechanism Behind the "Same Shape" Theory

Someone may notice that no student articulated a mechanism for why the bubble should come out the same shape as the wand. We can try to fill that in: What sense of mechanism would support that answer? When would that reasoning be correct? It's not hard to think of situations, such as pushing dough through a cookie mold: it comes out the shape of the mold. If that's what students were thinking, it might be why they didn't explain the mechanism: *of course* it would come out the same shape—how else could it get through the hole? That may be what was behind Jasmine's comment in line 64, "Why would it come through in a circle if it's a triangle?"

Staying Respectful of the Students

There's one thing further to say here. Most groups find Lester's comment about creating a rectangular bubble a little amusing—we can be quite sure Lester had never tried it at home and blown a rectangle. But that amusement over human nature should not become disrespect for this student. It is not uncommon that children support their answers by saying that they have tried it or seen it when we have reason to doubt they have.

One interpretation, of course, is that the child is simply making it up: he hasn't really tried it, but he expects that saying he did will get others to take him more seriously. We try to model empathy for that: We all want to be taken seriously! And it's hard to find anyone who hasn't done something similar, at one point or another. In fact, Galileo is known to have done the same thing in writing: he reported experimental results historians have demonstrated he could not have obtained.

Another interpretation is that the child may honestly remember it as he has reported; he may actually believe he once blew a rectangular bubble. This is something else we all share: we sometimes remember things the way we think they should have happened rather than the way they did.

The Discussion to Line 198

We stop the video around line 198, about six minutes since the last stop, because there have been several bits we believe deserve attention, and they take some time to discuss. We start as usual with an open question to ask what people have noticed.

People sometimes start out with comments about how the children had started to speak with each other. Noticing the tenor of respectful disagreement, some say the students must have had good training in how to have discussions. (That may have happened in the school in general, but not this year: this discussion took place on September 4.)

Samantha's Declaration

People reliably notice Samantha's provocative declaration that even if she tried blowing a bubble and it came out the shape of the wand, she "still wouldn't believe it" (lines 102–3). We ask what participants think about that; is this progress toward or away from science? She was showing strong conviction for what made sense to her, but she was announcing that she would ignore empirical evidence! Talking about this moment, we focus first on making sure that both arguments come out—the need to respect empirical evidence on the one hand but the need to respect sense making on the other. And then there's another question: How should the fact that she's correct affect our interpretation?

This relates again to how people think about the roles of theory and evidence in science. In discussions that land too quickly on the view that Samantha should not take that stand toward evidence, we sometimes mention Einstein's famous declaration, on the eve of an empirical test of general relativity, that the theory was "too beautiful to be false," saying in effect that he would not believe evidence to the contrary. (The reverse has never happened—no group has thought it would be OK to ignore evidence in favor of intuition.)

The teacher asked Samantha to explain, and, as Zoë had earlier, she struggled to articulate why she thought corners weren't possible in a bubble. That makes a good focus for a moment: What might Samantha have meant, saying "it'd be silly" and "if you tried to touch the corners it'd pop"? Taken literally, the last part sounds as though she thought the bubble could have corners but you couldn't touch it; and of course touching any bubble usually makes it pop. One person suggested she was thinking the corners would be sharp, and sharp things pop bubbles.

Responses to Samantha

Zoë, the student who had earlier tried to express her sense of how the bubble formed, spoke up to say she thought she knew what Samantha meant (lines 115–22). This comes up in seminars as an example of how students were listening to each other, although some feel that Zoë wasn't successful at rephrasing Samantha's meaning.

People often miss Jerome's contribution (lines 129–33), perhaps distracted by the dubbing. We might point it out and ask, "What was Jerome thinking?" He said that the bubble from a rectangular wand would come out in a rectangle but it would pop right away. That is a different conjecture from Zoë's that the bubble would come out round. Maybe he was being more consistent with Samantha's idea that the corners would make the bubble pop? Is this a reasonable compromise between the cookie-dough mechanism of the bubble emerging in the shape of the wand and the idea that bubbles have to be round? And was this answer a result of an ambiguity in the question? Were students unsure whether the teacher was asking about the shape of the bubble *as it came out* of the wand or about the shape it would end up taking?

Samuel's Oval

We point out Samuel's idea (lines 155–57) about an oval, unless someone else does, and ask for comments. Some people talk about him as having been listening to Samantha's and Zoë's thoughts about corners. He came up with a shape that was different from a circle but that didn't have corners, teasing apart two different aspects of the problem, offering a way to control for *corners* while still varying *shape*. For us, that's the main reason to focus on his idea. So many curricula are focused on helping children learn to control variables; here is a moment in which a child is starting to do it on his own.

How Might a Teacher Respond?

Depending on the interpretations, people might have a range of ideas for how the teacher might proceed. In the snippet, the teacher facilitated the discussion, often prompting students to explain or clarify their reasoning (lines 104, 108, 123–24, etc.); in some cases she tried to help express a student's idea (lines 147–49, 162–63, etc.). The group might think about her perceptions and interpretations—what was she seeing that led her to respond as she did?

That usually leads to a conversation about the children as trying but having difficulty putting their ideas into words; that could frame an instructional objective, to help children learn to be more descriptive and precise in expressing themselves. Following are some other possibilities teachers have raised for helping the children become more articulate:

- having students take a moment to write or draw their ideas, which would also provide another kind of opportunity for them to engage
- creating a list of ideas on the board, with the teacher writing and perhaps helping to edit the children's words
- pausing to check with other students regarding their understanding of what has been said, perhaps asking them to reexpress what they have heard in their own words.

The group should also, hopefully, have discussed what ideas about mechanism the children might have been trying to express. Responding to these interpretations, a teacher could try to help children identify and describe these mechanisms, such as by offering the comparison to a cookie press and asking if the children saw this as a similar thing to blowing bubbles. Some have suggested prompting children to think about their experiences with balloons in comparison to bubbles.

As always, every group will come up with its own set of ideas for instruction. There are no particular ones we raise if the group hasn't, but we work to be sure that (1) there are multiple possibilities seen as reasonable, each with its own advantages and liabilities and (2) the ideas for instruction are explicitly tied to interpretations of children's reasoning, referring in some way to what has happened so far in the snippet.

 ### The Rest of the Snippet

Usually we've had only enough time to discuss the snippet through line 197 in a one- to one-and-a-half-hour seminar, but there's plenty more to consider in the rest of the video. In seminars that will be reading and discussing Pat's case study, we make sure to let the video play through her exchange with Rondell, because she focused on it.

Kedra's Comments About Corners

Jerome (lines 199–203) reiterated his thinking that bubbles would come out of rectangular wands with rectangular shapes, but they would "pop fast" because of the corners; if the group hasn't talked about his idea already, this is another chance. The teacher pressed him to explain the mechanism—why would having corners make it pop (lines 206–7)? Kedra responded with the idea that "the corners will be like, get sharp and pop it" (line 208) and then elaborated (after the teacher asked her to "repeat that"), "the corners will, like, get a little, like, hole in it or something" (lines 210–11). This would be a good place to stop: "What was Kedra thinking?"

Rondell's Thinking About Changing Shape

Kevietta remembered a television show when bubbles came out the shape of the wand (lines 219–21), and the teacher framed that as another kind of experience. Someone might remark that this is good scientific behavior, to raise counterevidence; it would be important to account for it if the group's conclusion was that bubbles would come out round.

Rondell disagreed (lines 224–26), and he tried to explain. It is difficult to understand what he was trying to say, though; the teacher took a moment to try to pin it down (lines 227–50). Was he thinking that the bubble would come out the shape of the wand and then pop, like Jerome? One bit of evidence for that interpretation is the way he said "if it . . . changes into a circle" (line 235). Was he saying that the process of changing would make it pop?

Another comment that has come up here, when we've had time to watch and discuss this part of the video, is that the teacher spent a lot of time with this one student. What might she have seen or heard in his thinking that led her to respond that way? (Pat discusses this in her case study.)

Jasmine's New Reasoning

Jasmine was the next person to speak after Rondell, to say that the bubble "won't go through" (lines 254–55). How does her thinking at this point compare with her thinking earlier in the discussion? When she argued that the bubble wouldn't go through, was she thinking of a mechanism like getting stuck, or was she disagreeing with Jerome and Rondell to say that it would pop while it was still in the wand? A little later, Samuel (line 326) started to talk about the bubble in the wand as being like a spider web; maybe he got this image from listening to Jasmine?

Louis' New Evidence

Finally, Louis (lines 340–42) introduced a new piece of evidence into the discussion, that bubbles come out of the circular wands, but the circular wands have little "sharp things" that for some reason don't pop the bubbles. If those do not pop the bubbles, he was asking, why would the corners on rectangular wands?

The discussion continued for a few more minutes after the clip ends on the DVD; we stopped it there mainly to save us more blurring and dubbing for Jerome, who had a fair amount more to say.

Pat's Case Study

We have always shown and discussed at least the first ten minutes of the video before having people read and discuss the rest of the case study; sometimes we've had time for more.

Attention to Student Thinking

Talking about the case study, we might ask about the exchange with Rondell, which Pat addresses in some detail. If the group hasn't talked about this exchange before, when watching the video, it might be a good idea to play that segment again. How does the evidence line up with Pat's interpretations—how she originally saw it or how she saw it much later?

There's another aspect of her interpretation to notice, too: she says that Rondell hadn't participated much in class discussions so far, and that was part of the reason she gave him so much time in this class. It's useful, too, to ask about her strategies and objectives in responding to him: What was she trying to do, given her interpretations of how he was thinking? Was she trying to persuade or dissuade him one way or the other, or was she only trying to help him clarify his reasoning?

Discussing "Should Haves" of Interpretations and Actions

The topic of missing things in class comes up in other places, too, in Pat's case study, and we often try to frame a conversation about that. Reviewing student work, especially when it is rich with ideas, is like rereading a good book or watching a good movie another time: There are always new things to notice. It doesn't mean we did anything wrong the first time around! It's just not possible to catch everything. But Pat's case study raises another useful question: In instructional practice, how should we take into account the fact that our interpretations can be mistaken?

Several of Pat's comments are about what she should have done in the class, or about an idea or reasoning she missed at the time, with the hindsight from having watched the discussion on videotape and discussed it with colleagues. For example, she writes that she "should have asked [Louis] why he thought they would all be circles" and "should have gone back to Jasmine's thought" and "should have asked [Zoë] more about what she meant." On the other hand, some readers notice, at other moments Pat questions whether she might have thrown a student off by asking questions—Zoë at one point and Rondell later. Since these are all moments accessible on the video, it might be interesting to watch one or more of those moments and consider the possibilities: What are likely interpretations of what was happening, and what are some items on the menu of instructional possibilities?

Trying a Science Talk for the First Time

It's important to note that Pat and her students were trying a science talk* for the first time. As happens with "The Pendulum Question" (Chapter 4), people often express surprise that this is the first time the teacher and students have held this sort of discussion. Of course, that's part of why we chose these cases for the collection, thinking they might be especially helpful for readers trying to do this sort of thing with their students.

Pat writes that, at the time, she was not happy with how the talk went, after having written about her uncertainty with respect to having the talk in the first place. It was only after watching the video later, with peers who helped her focus on what the students' were saying and doing, that she decided it wasn't such a bad class after all. It took this reexamination of the data for her to appreciate how well the students were doing, including several who had not spoken much in class before. Both the anticipation and the assessment can be relevant topics for seminar participants who are contemplating (or not) whether to give their students this sort of opportunity to express themselves. Perhaps their students are capable of more than they expect; perhaps, too, they should not be too hasty to decide a discussion didn't go well.

---

* We sometimes remind participants of how this term came from Karen Gallas' book and how in the project it came to mean any conversation about science in which students were free to express their own reasoning.

CHAPTER 8

Second Graders Discuss Magnets

Kathy Swire is a science and mathematics specialist who travels around Frederick County, Maryland, visiting classes. She joined the project in the second year, so she'd missed the conversations we'd had the first year about why magnetism isn't a great topic for elementary school.

We'd seen several snippets of children talking about magnets, which was the topic of one of the kits several teachers were using, but we saw little in the way of tangible, mechanistic reasoning. We'd come to the conclusion that magnetism just wasn't a good subject: What resources would children have for thinking about how magnets work? So we were all ready to tell Kathy about what we'd learned. But her snippet got us thinking again about what was possible. That's one of the reasons we chose to use it: Whatever our expectations, we should look at the evidence and see whether they hold. Sometimes they don't!

The students in this case study had been working with magnets with their teacher; as a science specialist, Kathy was not with the students every day. So she began the class by asking them to tell her about what they'd been doing. Along the way, she heard something that excited the students and that she thought might make for a productive discussion. The video starts with the class discussion already under way; read the introduction from her case (see page 135) for more information.

Our experience in workshops and seminars makes us expect this is a more difficult case for people to interpret for the beginnings of science in the students' thinking. So we have not used it as a first video in seminars. (Of course, we could be wrong.) The transcript is available on the DVD, and the facilitators' notes come after the case study.

 ## Suggestions for First Viewing

The video on the DVD is a little less than fifteen minutes long, and it's possible to watch all of it in a one- to one-and-a-half-hour session. Here are some good stopping points:

1. *While the students are discussing the question in groups, about a minute into the video (if you stop at line 24, you can start the DVD again with "These two said . . ."). We stop here to talk briefly about the phenomenon the children have discovered, which Kathy has chosen as a focus for discussion.*
2. *After Kathy asks Craig what he was thinking, a little less than six minutes into the video (lines 105–6).* What did you notice about the children's reasoning?
3. *After Savannah says, "Sometimes," a little more than nine minutes in (line 178).*
4. *The end of the snippet.*

■ ■ ■

The Power of Magnets

Kathy Swire, Frederick County Public Schools, Maryland

I remember playing as a young child with two magnets shaped liked dogs at my grandmother's house and being fascinated when the two ends would pull together or push apart. The second graders I worked with this year had the same fascination.

I'm a science and math facilitator for Frederick County Public Schools in Maryland. My job entails working with teachers in schools around the county to support their teaching of science and math in elementary school. At the beginning of the school year I mainly work with new teachers, and that's how I met and started working with this class in one of the county's small, rural elementary schools. There were sixteen students in the class.

It was February, and the class had been working with magnets as part of the county's physical science unit, "Interactions and Systems." The teacher had told me how the students were excited about the topic and how they'd been testing items in the room to see what would stick and bringing in things to test from home. Planning the lesson, I decided to have them tell me about what they'd been doing and then ask them why they thought magnets sometimes stuck together and sometimes pushed apart.

I started by asking the students to tell me what they knew about magnets, and they were very eager. And then, as frequently happens, the students changed my plan! Dalton and Craig shared something they'd noticed in exploring two large magnets: When Craig had one magnet on top of his hand, with his palm facing down, he could put a second magnet on his palm and it would stick there. It even moved in response to the top magnet being moved. The students were so obviously thrilled by this discovery that I wanted to help them delve into it a little.

So I asked them to talk about why the bottom magnet would stick and why it moved when the top one moved. I chose to have the students first discuss their ideas in small groups. That would give all the students a chance to put ideas out on the table, so to speak, and to solidify what they had to say, which might help them share with the whole class. It also allowed me to move among groups and hear their ideas.

Maybe "We Have Metal in Our Bodies"? (Scene 2)

When I brought the class back together, I called on Calvin first, who normally seemed distracted during our conversations. He told everyone about something he'd heard from two other students, who thought the reason the magnet stuck to Craig's hand was that "we have metal in our bodies." That matched with an earlier part of our discussion, when the students said magnets stick to certain kinds of metal, but Calvin didn't agree with it. He said that we don't have metal in our bodies, that "the hard things is our bones."

Savannah spoke up to say that she had a piece of metal in her hand, which was why she thought it worked on her, but she didn't "know about everyone else." Then Taylor added a piece of evidence from her experience (lines 55–57).

> **Taylor:** Because I got to take a trip to a hospital before and I had to go in this room and wait for my cousin and there was an x-ray up on the board that I saw that—
>
> **Teacher:** So, you saw an x-ray of somebody who had metal in their body?
>
> **Ben:** He was probably wearing something like I, um, I broke—I broke the [*It's hard to hear what he said, but he pointed to his collarbone.*] and I was wearing something that had metal, and you could see something, like, metalish, so he probably had something on.
>
> **Teacher:** Ah, so you think that person was wearing something—it wasn't actually in his body or her body?
>
> **Ben:** Yeah, it was.

Ben was refuting Taylor's evidence by explaining what she may have seen in the x-ray. Then Evan gave another reason not to believe there is metal in people's hands.

> **Evan:** I was thinking, I disagree with Savannah because if there was [a] metal piece in your hands then it would stick.
>
> **Teacher:** Oh! So, if there's metal in my hand, the metal should stick to my hand.

I wanted to help Evan clarify his idea, so I held out my hand with one magnet on top.

> **Teacher:** So, if I flip my hand over, what will happen?

Evan: It won't stick.

Teacher: OK, if there's metal in my hand and I turn my hand over, what should happen?

Evan: Um, it should stick.

Teacher: It should stick, OK. And—and why?

Evan: Um, because she said there's metal in your hand.

Teacher: OK, and you told me earlier that magnet does what to metal?

Evan: Um . . .

Teacher: What happens when you put metal close to magnet—magnet close to metal?

Evan: It sticks. (see line 104)

Evan seemed to have been mulling over what he knew about magnets and realized that Savannah's idea didn't match: if the reason the magnet stuck was that we have metal in our hands, then a single magnet should stick by itself, without the other magnet on top. He needed some help expressing that logic, but I was excited to hear his reasoning. I remember during the lesson thinking, "Yes! Yes! He's listening and connecting to other ideas! Wow! I have a second grader going through this process!"

Looking back, I realize it wasn't just Evan. Taylor had provided evidence to support Savannah's idea; Ben had listened to both Savannah's and Taylor's ideas and provided a possible explanation for what Taylor had seen. Second graders listening to and supporting or refuting each other's ideas was amazing.

The discussion went on to other topics, but before I get to them I'll skip to the discussion (Scene 4) when the students had more to say about the idea of metal in our hands. Calvin brought it up again, reiterating Evan's reasoning.

Calvin: You put the metal piece right here [under your hand], and you don't hold onto it, if it sticks then there's a piece of metal on your body.

Teacher: So, if I put metal here and a magnet here [*showing the top and bottom of her hand*]—

Calvin: Yes. No, no magnet there, but, um, he means put—I mean put the magnet right here. And don't hold onto it.

I did as he said, putting the magnet under my hand and then letting go, and the magnet fell off.

Calvin: There's probably no whole metal in your body.

Teacher: So, are you agreeing with what Evan said earlier?

Calvin: Yes.

Teacher: OK, because Evan said if there's metal in my hand and I turn it over, it should stick. You're saying, am I right, Calvin, you're saying if I put the magnet here and I have metal in my hand, it should stick to my hand?

Calvin: I mean metal in your body.

Teacher: In my body. Then this magnet should stick to my hand without me holding onto it.

Calvin: Yeah.

When I asked Calvin why that would happen, he got a little stuck and asked for help. I called on Savannah, who was the one who had originally offered the metal-in-the-body idea. It was interesting to listen to her restate Calvin's idea and then give a reason not to believe it.

Savannah: I think he's thinking that if he has metal in your hand, um, then you will be able to have a magnet stick to your hand without you holding onto it.

Teacher: OK, why though? Why does that work?

Savannah: Because—or sometimes if it's a wrong—it's a wrong kind of metal, it won't stick to your hand.

Teacher: The wrong kind of metal where?

Savannah: In your—or whatever part of your body.

Teacher: OK, so I have to have the right kind of metal in my body and then it would stick.

Savannah: For it to stick.

Teacher: So, are you saying I could still have metal in my body, but it's not going to stick because it's the wrong kind of metal? OK.

Savannah: Sometimes.

Savannah's idea of the "wrong kind" of metal connected again to earlier class work when the students had discovered that magnets would not stick to all kinds of metal. She was using that to refute Evan's and Calvin's method of testing for the presence of metal: that piece in her hand could have been the wrong kind.

"It Has to Go Through Your Hand"
The second possibility for explaining how the bottom magnet stuck to Craig's hand came from Dalton. He suggested, after Taylor's x-ray comments, that the magnet had some kind of power.

Dalton: Um, that if you—if, um, the magnet moves, then, um, on the bottom, then it has to go through your—through your hand and, um, but it doesn't hurt and it moves the magnet. It has some of the—some power in it. (See line 67.)

Teacher: So, power in it. OK, and you told me that it—the power was like . . .

Dalton: Um, kinda like electricity.

Teacher: Kind of like electricity. But it's—is it electricity?

Dalton: No.

Teacher: No, OK, but . . .

Dalton: It's—it's certain kind of electricity.

Teacher: OK, and it can go through my hand and it doesn't hurt me but it will move this magnet over here.

Dalton: Yeah, it will go.

I thought Dalton's idea would spark more conversation among the students, this idea of a power that didn't hurt and yet was strong enough to keep another magnet from falling off my hand. At the time, I remember thinking that it was a great thought and I was hoping the next student would go with that, but that's when Evan offered *his* idea about why we don't have metal in our bodies. I let Dalton's idea go, but Katie brought it up again later in Scene 3.

Teacher: OK, but how'd—how'd this bottom one move? If I just did this and the bottom one moved, how did that—

Katie: Yeah, but it goes through your hand.

Teacher: What does?

Katie: The magnet.

Teacher: The magnet goes through my hand?

Students: [*Laughter*]

Class: No!

Teacher: I'm just checking. I'm just checking. I'm getting a little nervous here. What's going through . . . ? 'Cause, wait a minute, does a magnet go through your hand?

Students: [*Laughter*]

Katie: No.

Teacher: OK. Now wait a minute, don't laugh, it's a good statement. I think I know where she's going, but I want her to explain a little bit more. Katie, tell me a little bit more about what you mean by the magnet "goes through" my hand.

Katie: 'Cause it doesn't go through your hand, it just, like, another big one just sticks on there and moves.

Teacher: [*Whispering*] Why?

Katie: I don't know. (End of Scene 3.)

Katie didn't mean the magnet went through my hand; I just wanted to nudge her to try to be more precise, and I knew I could joke with her about it. Toward the end of the talk, Ben came back to the idea of electricity, but it's hard to say what he or other students meant by that.

I felt like the students were just on the brink of expressing the mystery here of the power that could pass through someone's hand without hurting it. I didn't expect them to *solve* the mystery, but I thought it would be wonderful if they could say more clearly what it was. I decided to come back the following week and continue our conversation.

Afterthoughts

I remember leaving the classroom absolutely bowled over by the conversation. I couldn't believe how well the students had listened to each other and been able to add their own ideas to reflect or refute what someone else had said. It was exciting to see the intensity and sophistication of their conversation. For a second-grade class, I was absolutely thrilled!

How did they come so far since the start of the year? I knew that my facilitating science talks on a biweekly basis had not generated this level of conversation by itself. In speaking with the classroom teacher, I found out she'd been using talks in most subject areas, having observed me in science since the start of the year. She felt that the students were getting pretty good at explaining themselves and adding to each other's ideas. That was exciting, too, to see a first-year teacher getting something out of my visits and applying it in her teaching.

In reflecting on the lesson with colleagues, someone asked why I hadn't demonstrated the two-magnets phenomenon. I wasn't sure why. I don't remember choosing not to use the two magnets. I actually had both in my pocket but chose to pull out only one to keep the students focused during our conversation. I think subconsciously I decided early in the conversation not to demonstrate; I wanted the students to think and reflect on their ideas but not be distracted by or want to try the experiment with the two magnets. None of the children seemed to care. No one asked to try it or see it done.

I came back the next week, and the children picked up where they'd left off, talking about the power that was "like electricity." Calvin shared a new experience in which he had a magnet stuck to a rock and kept three paper clips hanging from the other side of the rock—the students talked about the "power" going through the rock and paper clips. Taylor introduced the idea that big magnets have little magnets inside them, and Dalton asked what was inside the little magnets!

They didn't resolve what exactly the power was, but they continued with a similar intensity and sophistication, listening to each other and supporting and refuting ideas with evidence and logic. I felt like I could have gone back again and continued this conversation, but I also wasn't upset about letting this go. I remember the classroom teacher asking me, "When do you tell them the right answer?" My response for this topic was to let them explore and try to find an answer themselves, and not to worry if they didn't. The students didn't seem to worry. They were willing just to think about and discuss their ideas.

It was interesting to me to find out later that several others in the project had held discussions about magnets with their classes, but the discussions hadn't been successful. I'm glad I didn't know that in advance or I might not have tried this talk, and I'd never have seen what can happen.

■ ■ ■

Facilitators' Notes

Please see the general notes for facilitators in Chapter 3. We don't intend these notes to be a thorough analysis of the snippet but rather to give a sense of some possibilities for topics that might arise and that a facilitator might choose to raise.

What Is There to Say About How the Magnets Hold Each Other?

It may be worth experiencing the magnet phenomenon in a seminar: the magnets stay on your hand because they're attracted to each other. In some groups, people talk about how magnets can also repel, and maybe about north and south poles to say that likes repel and opposites attract. That's as much as we can do—talk about the rules magnets obey but not go further to understand why those are the rules or how magnetism works. A few people may talk about magnets being made up of little magnets; still, as Kathy's student Dalton asked, what makes the little magnets magnets?

As an aside, so people don't feel bad, it might be worth saying something about how this was a very difficult puzzle in the history of science. (In one early model scientists envisioned lots of tiny spinning particles shaped like right-handed or left-handed screws that, when they touched each other, would either pull toward each other or push away. It's a nice example of how scientists try to use tangible mechanisms to explain phenomena.)

When we ask participants what they think the children might say about why the magnets stuck to a child's hand, they have as hard a time as we do coming up with ideas. Children can see the magnets attract some things and not others, and maybe they find out that magnets attract only some kinds of metals. But we can't think of any everyday resources they could use to understand the mechanism of attraction or what makes some materials respond to magnets but not others. There's nothing else they know that's like magnetism. Without any mechanistic options, it seems easy for children to fall into other sorts of thinking, of nonmechanistic storytelling.

When we've used this case in seminars, we've always talked about our prior experience in the project with magnets as a topic. That's probably why we haven't heard participants raise their objections: we've raised them first! We give the background that magnets were part of the county curriculum—neither Kathy's choice nor ours. But then the focus should return to the students: What did these particular children do, and is it in line with what we anticipated?

The video skips the children working in groups, a choice we made for several reasons. One was that the room was small enough and the children loud enough that the microphones all picked up several groups speaking at once. Another was that we felt there was enough to see and talk about in the full-class discussion that followed.

The Discussion to Line 106

Opening the Conversation

Line 106 is a good place to stop because the students have raised a couple of ideas and batted them around a bit. Invariably what stands out is the idea Savannah and Taylor raised, that people have metal in their bodies. People also tend to notice and appreciate the way different children were bringing up evidence for and against the idea, without any prompting.

The Idea That We Have Metal in Our Bodies

It is worth taking stock of the different arguments the children raised; we ask people to do this, citing lines in the transcript: What reasons did the children have for believing there is metal in our bodies, and what reasons did they have to disbelieve it?

It's easy to find their reasons: Savannah's idea about how she had a piece of metal stuck in her hand (lines 47–49) and Taylor's story about her trip to the hospital (lines 55–57). Were they each saying that all people have metal in their hands, or only some people?

Ben's response to Taylor (lines 60–62) should come up in the conversation. If no one else does, we point out that he was not contradicting what Taylor saw but explaining it in a different way, a simple reconciliation of her experience with his belief that people do not have metal in their bodies.

Evan's response should come up as well (lines 90–104). Some people may have the reaction that Kathy was putting words in his mouth, which makes this a moment to talk about interpretation and response: What do participants think Evan was trying to say? What did Kathy think Evan was trying to say? The sense that Kathy was putting words in his mouth is, in essence, a disagreement of interpretation regarding what he was thinking. We also ask, "What need did Kathy see, on Evan's part, that she was trying to address?"

Another place perhaps to pause and interpret is Dalton's point that "you'd have to be at the doctor to do it" (line 82)—to do what? Surgery to implant metal? (Some may enjoy the *Twilight Zone* reference!)

The Idea of "Going Through Your Hand"

The other idea people may notice is Dalton's thought about the magnetic "power" (line 69) being "kinda like electricity" (line 71).

In most seminars for elementary educators, the participants will not have a clear sense for themselves of the difference between electricity and magnetism, and noticing this moment is likely to either make people uncomfortable because they do not know whether magnetism is a kind of electricity or, in a strong group, inspire a digression to talk about it.

If people seem uncomfortable, it can help simply to make that explicit: "We might not be in a position to decide whether Dalton is right or not, and that's fine; can we talk about what makes him say that?" But if they want to explore the physics topic for a little bit, that could be worth a digression. One useful question might be this: "Can you use magnets instead of batteries in a flashlight?" Our approach has been to let that conversation happen for a bit and then return to focus on the students. It's important that people not feel disqualified from reflecting on the students' reasoning because they don't know the answer themselves. We haven't found that to be a problem, maybe because we use this case study after several others, when the group has come to understand the game.

Few people notice what Dalton said before referring to electricity, but we see it as an important bit, so if nobody else mentions it, we point it out: "It has to go through your hand . . . , but it doesn't hurt" (lines 67–68). Why does he say "but it doesn't hurt"? The idea of something going through your hand comes up again later, too.

The Discussion to Line 180

It's good to stop at line 180 because students have said things worth thinking about, and the next student to speak is going to change the topic. The question of whether people have metal in their bodies came up again, as did the idea of something "going through" your hand, from one magnet to the other.

Return to Metal-in-the-Body Idea

For more than half of the segment the students focused on the question of whether people have metal in their bodies. Most people will notice Calvin's explanation that if there were metal in someone's body, one magnet could stick by itself (lines 142–61). Was there any difference between Calvin's thinking here and Evan's earlier, or was he essentially saying the same thing?

What do people see in Savannah's reasoning (lines 166–78)? She volunteered to say what she thought Calvin was thinking, but she went on to express an objection to it: if the metal were the "wrong kind," the magnet still wouldn't stick (lines 169–75). If nobody remembers, it may be helpful to remind participants that Savannah was the one who said she had a piece of metal in her hand. (If it was the wrong kind of metal, that would account for why a magnet wouldn't stick to her hand.)

What Goes Through Your Hand?
Katie expressed the idea that "it goes through your hand" (line 117), and the exchange that followed stands out in the video. This is an opportunity to talk about interpretation and response: What was Kathy's interpretation of Katie's thinking that caused her to respond as she did? No one will believe Kathy thought Katie meant the magnet would go through her hand; it's easy to recognize the strategy of taking a student literally to prompt her to be more precise.

The problem, of course, is how to be precise. Depending on what came up in the conversation about Dalton's similar comment in the previous segment, it might be useful to linger on the question for a moment and maybe compare students' idea of something "going through" the hand to the seminar participants' ideas about magnetism: Do they also have a sense of something moving or passing between the magnets? Another way to think about it is that the two magnets interact with each other directly, so nothing passes between them, so nothing goes through anyone's hand. (Physicists call that idea "action at a distance." See the Notes for a little more.)

The comments from Katie and Dalton relate to the question of whether magnetism is an appropriate topic for children to explore. At some point in the conversation we want to talk about that. This is one opportunity; there will be another in the last part of the discussion, and another when talking about Kathy's case study.

On one hand is the view we'd formed in the project, that children just do not have any sense of mechanism about why magnets attract or repel; there's nothing else in their experience they can use to relate to that. So how could they have tangible ideas for what causes magnetism? On the other hand is the interpretation that these children were *trying* to find a tangible explanation. So they had a sense that *something* "goes through you hand," but that something was strange to them. In that, they were finding questions physicists have wrestled over, and finding questions is a kind of progress we should value.

The Rest of the Snippet
There are several places it could be useful to pause and interpret student thinking in the final segment. A few ideas came up that the students did not pursue because they were running out of time.

One idea that came up briefly is that it may have an effect on how a magnet sticks if there is plastic covering the metal. Dalton (lines 181–87) was referring to comments Evan had made in their small group. Renee (lines 233–38) seemed to have heard Savannah say something about thin or thick plastic in their group.

Taylor came back to the idea that it had to be the right kind of metal (lines 200–202) in order for the magnet to stick. When she talked about the "backwards side" of the magnet (lines 200–204), was she thinking that one side of a magnet attracted metal and the other repelled it?

Ben came back to the idea of electricity going "through our hand" (lines 218–21); what might he have understood *electricity* to mean? And what might Calvin have been thinking was wrong with that statement, in his disagreement (lines 223–27)?

A useful topic for a seminar conversation might be to assess where the students seemed to be, as a group or individually, and from there consider the menu of possibilities for how the teacher might start things next time.

Kathy's Case Study

As usual, we start the conversation about the case study by asking what people have to say. There are several good topics, including Kathy's remark about changing her plans, and her objectives at the start and then at the end for talking about magnets. Some might comment about how Kathy's interpretations did or didn't align with what they had thought when watching the video. There are a few topics we watch for and might raise if nobody else does.

Changing the Plans

Kathy wrote that at the beginning she was expecting to ask children about what made magnets sometimes attract and sometimes repel. What might we expect would be within the children's abilities? Most people think they could understand the idea of magnets having two kinds of poles, north and south, and that north and south attract, but like poles (north-north or south-south) repel. They could try this out, collect data systematically, and show that's what was happening. And many people think it would be valuable for the children to see these aspects of magnetism.

But, Kathy wrote, she changed her plan for the class based on what she heard. The students had found a phenomenon that intrigued and puzzled them, and they were excited to talk about it. If the seminar is made up of experienced teachers, they will likely have a lot to say about this, perhaps offering examples from their own practice, and this is a useful conversation. Depending on their particular circumstances, some may have things to say about scripted lessons or mandated daily objectives, and that would be a useful conversation to have as well.

We have a point of view about this, of course: our purpose is to help teachers learn to hear and respond to their students' thinking. When running seminar conversations, we do not pretend to be neutral! If teachers are not free to hear and respond to their students' ideas and interests, then students are likely to learn that their ideas and interests are not important in school.

Interpretations and Responses

Talking about the discussion, Kathy had particular things to say about student comments that might or might not align with a group's reactions when watching the video. People tend to notice and talk about this for themselves; otherwise we ask, "How did Kathy's interpretations of the students' thinking compare with ours?"

For example, one place to focus attention could be on how Kathy remembered thinking, "Wow, I have a second grader going through this process," after Evan gave his reasoning about metal in our bodies. What was she seeing as "this process" in his thinking? And what more did she recognize later, looking back at the class, about the other students?

We'd like to be sure people consider particular interpretations, including Evan's reasoning as reconciling Taylor's evidence, accounting as he did for what she had seen in the x-rays, and Savannah's reconciling why a magnet couldn't stick to her hand by itself when she had a piece of metal stuck inside.

For any of these instances, if the group's interpretations are in line with Kathy's, it could be useful to think about possible ways to respond—not what Kathy *should* have done, but other items on the menu of possibilities. Of course, if the group's interpretations are different from hers, there's all the more room for that conversation.

The Mysterious Power

Kathy also wrote about Dalton's idea about power as a "great thought" that she was hoping other students would take up. If the seminar has not already talked about this theme of the case, this would be the moment to raise it: What did she see as great about Dalton's thought? It's not important that participants agree with Kathy's assessment; we're looking only for them to understand it. Given her interpretation, how did she respond, and what are some other possibilities?

A more general topic of conversation that might arise concerns what science instruction should hope to accomplish. Here, Kathy was talking about trying to get students to understand a *question*, one that concerned their sense of mechanism. Do people see that as an appropriate objective? It's not a standard one, of course, in curricula or in state frameworks. Again, we do not pretend to be neutral: We think it should be seen as a valuable objective and accomplishment; certainly it is in professional science, where articulating a question can be a pivotal contribution. But we would not hope or expect most seminars to arrive at a consensus with us; our purpose is that participants understand the arguments on either side. (Which is to say, of course, that our objective in these conversations is to raise the question!)

CHAPTER 9

Eighth Graders Discuss Chemical and Physical Change

Steve Longenecker had never been happy with the results of teaching chemical and physical change. As Jessica Phelan had done with the rock cycle ("Chaos in the Corridor," Chapter 6), Steve tried a new approach, asking the students to come up with their own ideas. But where Jessica found the students could find their own ways toward the conceptual objective in the geology unit, Steve found a conflict: given what they knew, if the students acted like good scientists, they should be dissatisfied with the definitions he was supposed to teach.

There's no video for this case study—we didn't have permissions—but we decided to include it here at the end to introduce a next step in this conversation: Having developed practices of hearing and responding to the substance of students' thinking, of recognizing and supporting the beginnings of scientific inquiry, how might teachers think about coordinating those practices with traditional objectives? And in particular, how might they do this when the traditional objectives involve concepts that are not accessible as refinements of everyday thinking?

■ ■ ■

Teaching Chemical and Physical Change

Steve Longenecker, Montgomery County Public Schools, Maryland

I taught eighth-grade science at a middle school in suburban Maryland just outside Washington, D.C. This was one of my smallest classes, with only nineteen students.

The school had recently switched to a block schedule, which meant that each of my classes met every other day for a ninety-minute period. Because the middle school curriculum was spiraling, each year students had three to four units on different science topics. The chemistry unit was supposed to be about seven weeks long. Although the students had certainly studied various aspects of physical science in elementary school (they were all terribly familiar with solids, liquids, and gases, for example), this unit was their first formal introduction to chemistry. Their expectations were high, and many of them (boys especially, it seemed) immediately wanted to get out fancy glassware and Bunsen burners and "start blowing things up."

The Problem with MLO 4.8.3

The objectives for this unit were determined at the county level, and they had been in some flux. First, new countywide objectives were established, and then the state of Maryland published its own learning objectives, which forced the district to reformulate its guidelines once again.

Throughout all this change, however, the overall approach at both the state and local levels stayed fairly consistent. Goals for student learning were usually grounded in the notion that ideas be generated and shaped by "concrete," observable phenomena. For instance, a typical objective was that students be able to "distinguish one substance from another based on observable and measurable properties (i.e., density, boiling, melting point)." This objective can be achieved through fairly straightforward experimentation and observation.

Noticeably absent from the state and district learning objectives were traditional expectations that students learn about the parts of atoms, how atoms interact to form molecules, and so forth. In fact, only one Maryland Learning Objective (MLO) even mentions the word *atom* at all (MLO 4.8.8: "Students will be able to explain that atoms and molecules are in constant motion and that an increase in temperature will increase that motion"), and the district did not include a corresponding objective in its local curriculum. This choice makes sense to me as an educator. The molecular and atomic models of matter are highly sophisticated, based on the experiences of generations of physical scientists, and treating them well is difficult at the undergraduate level, let alone with eighth graders. At best, students understand the rules of the model and its tight internal logic but make little connection to real life. (How can the basic particle of matter be mostly empty space? What is this electromagnetic force, really? How are electrons shared, really?) At worst, students find the model makes little sense and resort to memorizing rules.

Nonetheless, vestiges of traditional chemistry pedagogy remain in some of the objectives the county asked me to teach. Teaching with the "inquiry method" that county administrators championed, while still meeting these more traditional learning objectives, was a challenge I sometimes didn't meet. Either the objective or the inquiry had to go. This case study describes my attempt to have it both ways with MLO 4.8.3, which states that students will be able to "distinguish between chemical and physical changes based on properties."

On the face of it, MLO 4.8.3 should not cause a teacher consternation. Chemical and physical changes are to be distinguished by properties, by the physically observable attributes of the matter before and after it has changed. However, the actual definitions of these two kinds of changes are based on molecular structure: a chemical change is a molecular change. In general, this isn't something we can observe directly.

Solid (ice) and liquid water certainly have different properties, but changing from one to the other is considered to be a physical change because the matter has not changed its chemical composition (it's still H_2O). Salt dissolving in water is also considered a physical change because the solute and solvent both still have the same chemical composition, but the properties of saltwater are quite different from the properties of salt and water. Drop an Alka-Seltzer tablet in water, and it's a chemical change because a new chemical, CO_2, is being formed, but how could students know that CO_2 wasn't in the tablet to begin with? It's probably a distinction that should be retired from textbooks. I had attempted to teach it for years and had never been happy with the results.

Redefining the Goal

Conversations with colleagues about this frustration sometimes led to the suggestion that I discuss my concern with the students. Thinking about how I could do this, I decided to focus my attention on the science skills of categorization and definition. I would let my students work through examples just as early chemists (presumably) had when they developed the chemical and physical change schema described in the learning objectives. But I would let my students develop their own schema. Making sure they ultimately knew how various changes were categorized in their textbook would remain a goal, but it would be secondary to teaching them about the process.

I realized that this particular case could prove particularly useful for teaching that categorization can be a little arbitrary and debatable. I also hoped that highlighting the arbitrariness of the categories would make it easier for students to accept the shades of gray to this particular categorization and therefore learn the definitions.

Getting Started: Students Create Categories

I asked the students to observe and describe twelve changes of matter (such as a candle burning and paper being cut up) in a lab activity. For homework, I asked students to classify the changes into two, three, or four categories with the rule that each of the changes had to fit into one and only one category. Interestingly, students created categories that worked for the twelve examples we had observed in the lab, but most of the students created categories that were not more generally mutually exclusive. For example, one student had created the categories "involves heat" and "involves more than two things." This worked because only one of the changes in our lab involved more than two things and it didn't produce a noticeable temperature change, but of course changes that involve more than two things as well as temperature changes do exist.

As a follow-up to the categorizing assignment, we made posters of categorization methods in class. However, I adjusted the assignment by adding three more changes to the list we would categorize and saying that more could be added at any time, so their system had to be able to work for all changes of matter. Some students struggled with this second version, and I helped them with ideas. In the end, the posters showed a nice range of possibilities, classifying changes in various ways, including (1) fast, medium, or slow; (2) natural or human-made; (3) involving only solids, only liquids, or different phases of matter; (4) producing heat or not; and (5) reversible ("you can change back") or irreversible.

In hindsight, I realize that the results of these activities illustrated an important aspect of good scientific categories that wasn't second nature to the students. When scientists make categories, they want their categories to fit all examples, even reasonable hypothetical examples. Although my activities probably suggested this aspect of scientific categorization, I could have taken the opportunity to lecture briefly on it explicitly, possibly using examples of other scientific categories with which students were familiar (living versus nonliving, for example).

Class Discussion: *Comparing Three Changes*

The next class period, I asked students to write down whether salt dissolving in water was more like the change of an Alka-Seltzer tablet fizzing in water, a change we had observed, or more like glass beads (marbles) being added to water. We then went over their ideas in a class discussion. I considered the discussion an opportunity to assess students' thinking about categorization methods and for them to learn that members of a set can be categorized in more than one way with logical integrity.

> **Andy:** I think that the change when you put Alka-Seltzer in water is more like salt than the beads, because when you put the Alka-Seltzer in and the salt in, both will disappear but the beads don't. At least I don't think glass beads are a soluble material . . .

> **Teacher:** I don't think the glass beads would disappear, at least not very fast.

> **Andy:** Yeah, and also both change the chemical composition of the water. They both put something else in, while the beads just . . . they're the only thing that change the water. Like the salt and Alka-Seltzer, they actually change what the water is, but all the beads do is put the [something] in it. Yeah—

> **Teacher:** Hmmm, that's interesting, isn't it? Do you think that [argument] actually works?

> **Andy:** Well, the Alka-Seltzer and the salt actually mixes with the water; the beads can't really . . .

Lamar: I'm saying that the beads take up a lot more matter than salt or Alka-Seltzer. They take up a lot more matter, a lot more space, volume, than the . . . I'm just saying what Andy is saying.

Teacher: No, I think you're saying something different. You're saying when you add the beads to the water, they take more space up in your container, but when you add the salt or the Alka-Seltzer to the water, it doesn't really change the amount of space that the water is taking up.

Lamar: Yeah.

Kathleen: I agree with them two. And also because if you put salt or Alka-Seltzer in, you couldn't easily just put your hand in and take them back out. But with glass beads you can just put your hand in and take them out.

Teacher: Reversibility?

Kathleen: Yeah.

I was pleased that so soon after the discussion began, students had proposed three different categorization methods. Andy suggested that the "solubility" of the Alka-Seltzer and salt made them different from the glass beads. I questioned him about that because I wanted students to be able to defend the logic of their proposals. He started to, but wasn't convinced himself, so he let the point drop. Since I wasn't focused on any one method but more on the idea that any method be logically sound, I allowed the idea to drop.

Lamar's suggestion was similar to Andy's, as he said, since the volume of the resulting liquid he considered was affected by the "solubility," but it nonetheless used a different criterion than Andy proposed. Interestingly, both students were using very concrete and observable criteria in their proposals. Although Andy used the term *soluble*, his categorization was actually determined by whether the substance added to water disappeared.

Kathleen proposed a categorization based on reversibility, which introduced a different level of abstraction. Instead of focusing on directly observable variables such as how fast the change occurred, Kathleen wanted us to consider a potential future of the change. I considered directing the discussion toward deeper analysis of her idea, which might have led students to think about the molecular nature of the matter undergoing change, as most irreversible changes are irreversible because the atoms have rearranged their molecular structures.

Indeed, many textbooks tell students that irreversibility is one way they can identify a change as chemical rather than physical. Of course, there are irreversible changes that don't involve molecular change (cutting paper into little pieces, for example) and I would have expected further student deliberation to produce these counterexamples. It might have been a rich conversation. But while I was tempted to lead the discussion in this direction, I chose not to, because I wanted the students to feel they were directing the discussion as much as I was.

Later, a student suggested an alternative take on Kathleen's idea of reversibility.

> **Ben:** OK, I disagree with *everybody* who said something because I think beads are actually more like the salt because . . . a few things. One, when Alka-Seltzer is dropped in water, it does *not* disintegrate into the water, it turns into a gas, I'm not sure which gas, and you can never get them back. With beads you can just stick your hand in and take them out. With salt, you can evaporate the water and get the salt back.

> **Teacher:** OK, that's interesting. You're kind of . . . I mean Kathleen said something very similar but she came to a different conclusion because for her, the evaporating is harder.

> **Ben:** Yeah, it's harder but it can be done. With Alka-Seltzer you'd actually have to grab molecules out of the air, and that we can't do.

I appreciated Ben's contribution in a number of ways. In finding the similarity between two phenomena that at face value are very different, Ben showed a fairly sophisticated understanding of the chemistry involved in all three changes. In fact, the chemistry textbooks agree with him that salt dissolving in water and glass beads being dropped into water are both physical changes; they argue that neither change affects the molecular structure of the materials (which is why they are both reversible, as Ben suggested). Ben's understanding of the reaction of Alka-Seltzer with water was also essentially correct, as atoms are rearranged in that process to form molecules of the gas carbon dioxide.

I also appreciated that while Ben disagreed with Kathleen, his argument was essentially an elaboration of her proposal to use reversibility as a guiding principle. After my earlier decision not to guide the discussion toward deeper analysis of this idea, I was pleased that it was proceeding in that direction anyway.

As the discussion unfolded, I didn't want to give more credence to either student's interpretation of what *reversibility* meant. Both proposals were logically sound. Scientifically accepted categorization methods often have areas in which one can argue about where to draw the line, and I liked how Ben's and Kathleen's alternate takes on the same issue illustrated this. Interestingly, the "where to draw the line" issue appeared again in the conversation later, when a student contended Andy's proposal for categorizing by solubility by suggesting that glass beads were also soluble, given enough time and enough water.

The Next Step: How Do Scientists Do It?

At this point, students seemed to understand that there are various ways to categorize changes to matter, and that as long as the categorization method is logically consistent and works for all changes one encounters, it is legitimate. At the same time, I had informed the students that scientists had agreed on one particular method of categorization for changes of matter. Students very much wanted me to tell them what this method was, which I attributed to the engaging quality of the earlier activities.

The next time we met, I cut out little strips of transparency film and wrote on them the changes that students had already categorized using their own methods. I put the strips into two columns on the overhead projector.

I didn't tell students the definitions of the columns; I just called them "left" and "right." (Chemical changes were on the left, physical on the right.) After I'd gotten about five of the changes categorized, I began asking students to tell me which column the remaining changes would go in. The class, as a group, was able to direct the changes to the proper column.

That conversation produced the two columns shown here:

- rusting of steel wool in vinegar
- baking soda's reaction with vinegar
- candle burning
- universal indicator turning blue when added to the base sodium hydroxide
- Alka-Seltzer fizzing in water
- flour being burned
- milk becoming sour

- mixing large and small paper clips
- cutting paper
- water warming up in hot-water bath
- mixing two different kinds of sand

With these lists on the overhead projector, I asked students to tell me what the definitions of the columns were. I said that I had heard Hal and Ben using the terms *chemical* and *nonchemical*, but I wanted to know what they meant and what was really going on.

Hal: I think that the left column is the chemical. Taking two things and putting them together to make something new, and . . . uh . . .

Teacher: What new are you making? Can we go then and say what we are making that's new? Alka-Seltzer fizzing in water?

Hal: You're making a gas. Universal indicator turning blue, uh, you're just making water changing color.

Teacher: How is that making something new?

Andy: You're making the universal indicator change colors, change chemical composition . . .

Teacher: Something in the universal indicator had to change in order for the color to change?

Andy: Probably.

Hal: Yeah.

Teacher: OK, I'll accept that. Flour being burned?

Hal: The flour is lighting on fire. It's changing.

Teacher: And what new are we getting?

Hal: It's like disappearing, I guess.

Andy: Gasses.

Kathleen: Smoke.

Lamar: Heat.

Teacher: There's some heat.

It was a pleasure to hear students proposing the "right" answer instead of telling it to them and asking them to remember it, particularly as the proposal came out of a process in which I felt that these students understood that this wasn't the only legitimate way to categorize these changes. However, I was somewhat concerned that the definition "making something new" was not especially sharp, a point brought home later in the conversation when Ben suggested that in every change something new is produced, but that the chemical changes are the changes in which it's new enough to be considered a different kind of matter. The challenge of this definition was made more clear when I came to the transparency strip with the change "salt being dissolved in water" written on it. Should it go in the left column or the right?

Andy: Left.

Teacher: Why left?

Andy: Because you're creating saltwater.

Teacher: So you're creating something new?

Andy: Right.

Joel: But if the [*unintelligible*].

Teacher: Something new that is different than something we've had before. Saltwater boils at a different temperature than regular water . . .

Andy: It's heavier.

Teacher: It's heavier. Salt is a solid, saltwater is only a liquid.

Andy: It might be a little more cloudy.

Joel: I disagree, because if you're mixing the two different kinds of sands, so you have black sand and white sand, you're mixing them together.

Teacher: Oh! So when you mix two different kinds of sands, you end up with a new thing, a mixture?

Joel: A new kind of sand. It's the same thing with the salt.

Mixing black and white sand together was a physical change according to scientists, and the students had been comfortable with the idea. Now, however, Joel was uncomfortable saying that dissolving salt in water was different from mixing the two sands. He and other students were exploring the logical extensions of these still-forming definitions and finding some of the weaknesses. This testing process is a key element of definition (or category) making and I was pleased to see the students making their arguments.

Later in the discussion, Betty said she disagreed with Joel that salt dissolving was the same as sand mixing. She made the distinction that the salt had actually changed form so that it was now part of a liquid, while the grains of sand hadn't changed at all.

Betty: They're just together.

Teacher: They're just together. But with salt being dissolved in water, you're saying, no, because . . .

Betty: Because it's *actually* being mixed together.

Teacher: Well, "mixed"? . . . You're trying to draw a line between "mixed somewhat" and "actually being mixed." Can you kind of define it more for me, what you mean? I'm not disagreeing with you, I'm just saying, where do you draw this line, actually being mixed and kind of being mixed?

Betty: I don't know. Um . . .

Teacher: Like, what's the difference for you? Why is Joel wrong that mixing two different kinds of sand together isn't different from salt being mixed in with the water?

Betty: I guess it could go either way, but there are, you would have to define [*unintelligible*].

Betty appeared to understand an idea about categorization and definition I wanted my students to see, that effective categorization demands sharp definitions.

I also wanted them to understand that categorization and definition are human efforts to impose order on a universe that isn't necessarily created according to categories or definitions. Sometimes we just live with anomalies; sometimes we respond to them by starting over with new categories. At the same time, attending to these larger issues may have distracted some students from thinking about the specific ways in which changes in matter can be categorized. I had to decide how much attention to give to each concern.

The arguments were summed up well by Ben and Hal at different points toward the end of the class period.

> **Ben:** Salt being dissolved in water should belong on the left, but mostly because the things on the right, they are not, they are not actually chemically bonded. If you're warming up water, it's just still water, nothing's really happening to it. If you're mixing up sands, there's no chemical bondage between the black grains of sand and the white grains of sand. You can still pry them apart with tweezers. You try prying apart dissolved salt and water with tweezers, you're going to get saltwater and saltwater cause they're actually bonded molecules.

Later, Hal added his idea:

> **Hal:** I think that when they are connecting, they're not making something new. Even though it is saltwater, I think it's kind of like salt and water. I think they are connecting, but it's just at a really small level. If there was some way, you could break them apart again. I might be wrong, I don't [know].

Eventually, I told the students that scientists (their textbook) considered salt dissolving in water to be a physical change (like mixing black and white sand together), but that I disagreed with this interpretation of the broader definitions. I explained that I could see both sides of the argument but found the reasoning that saltwater is a new kind of matter more compelling than the one that the two substances were just very well mixed. I also told them that I expected them to learn the idea of the textbook's categories and that they might just have to memorize the special cases of solubility and phase changes, since the logic of their categorization was so debatable.

I lectured a bit on categorization in general, making explicit my ideas about the imperfection of human definitions and ending with a brief discussion of duck-billed platypuses as another familiar real-life example that does not fit well into human-made categories.

Reflections

In many ways I was happy with this class. The students seemed less confused by the idea than students of previous years, and their work on the next assignment gave some evidence of their learning. I asked them to pretend they were eighteenth-century scientists and write a letter to their colleagues at the National Academy proposing a new method of categorizing the changes that matter undergoes. They had to address how dissolving would fit into their proposed method. Most of the students did well, providing definitions, examples, and arguments.

However, the discussion did not work for everyone. Some of the students who didn't participate understood the points being made and simply didn't want to talk, but others were silent because they did not follow the arguments. It might have helped to take breaks for students to do some independent writing or to have some of the discussion take place in small groups that could then report out to the class as a whole. Then every individual would have had more opportunity to speak up with ideas or questions and would have been more responsible for making an intellectual investment.

An area of categorization and definition that we didn't address was the evaluation of different schemes. Why are some methods better than others? Why do we traditionally categorize animals by their evolutionary relationship rather than by their habitat? (And then sometimes we categorize animals by their ecological relationship: carnivores, herbivores, etc.) It would have been interesting to talk about implied values (the molecular level seems to be more important to scientists than color, for example, which might be more important to a visual artist who was categorizing changes in matter), about scientific ideas such as *elegance*, and so on. These things did come up in the writing prompt, though, as I asked students to argue that their new method of categorizing was better than competing methods that the National Academy may have been considering.

In the end, I did go back and give students traditional handouts with definitions and examples of physical and chemical changes, and I expected students to know these definitions. This was in keeping with MLO 4.8.3 and with my urge (created by years of traditional science education?) to teach traditional science content. In some ways, though, I wish I had limited my goal simply to teaching about categorization and definition, without worrying about the "right" answer at all. My students would not be less educated if I had, especially given that the "right" answer isn't all that right, anyway.

Still, I do not regret telling them the textbook's answer, since I also told them I disagreed with it. Indeed, this may have provided a larger lesson. Textbook science is generally the explanation that most members of the scientific community agree upon. Students who work only through textbook science never learn how much science is argued quite vigorously. Studies disagree and researchers take different points of view. The discussions my students had about categorization provided them an example of this probably more interesting part of science. That's a lesson I also could have made more explicit.

Finally, I didn't think of it at the time, but an interesting enrichment for some of the students would have been to learn that the textbook categorization method was ultimately amended with the discovery of a new kind of change in matter: nuclear change. I wouldn't have pursued the tangent too far, but it would have been yet another illustration of the fluidity of scientific definitions and categories.

■ ■ ■

Facilitators' Notes

With all of the earlier case studies, the principal topic was always what we saw and heard in the students' reasoning, and we worked to defer conversations about what the teacher was doing. Having interpretations of the substance of their thinking, we could then consider possibilities for how a teacher might respond.

This time we move more directly to the latter, to talk about the possibilities that come out of different interpretations of the students' thinking. Steve did that a lot, reflecting on how he responded to what he saw and on what other options he considered or could see in retrospect. It can still take some facilitation to keep the conversation focused on instructional choices based on what these students were doing, rather than on students in the abstract. We try to do that, and, as always, we try to maintain an appropriate tone of respect for the teacher and for the challenges of teaching.

As always in these notes, we're not at all trying to be exhaustive about possible or required topics; these are some thoughts about what could come up and what a facilitator might choose to raise.

Preparation

In advance of assigning this case study, we generally pose the following question to seminar participants.

> Think of two kinds of change: (1) putting a bunch of marbles in water and (2) putting an Alka-Seltzer tablet in water. [It can't hurt to describe what happens: The tablet fizzes—bubbles form around it and rise to the surface—and disintegrates.]
>
> Now think of putting salt in water. Is that more like the first change or the second?

People reliably come up with a set of arguments comparable with what Steve reported in the case study. For some groups, the Alka-Seltzer tablet is the more natural comparison, because like dissolving salt, the tablet disappears (dissolves?) into the water, but of course everyone knows that if we're asking the question, there must be more to it than that. So they look for other ways of thinking about it, and maybe they come up with ideas about mixing rather than reacting. And in some groups there are people who know the topic well, name it "chemical and physical change," and can rattle off both sides of the argument along with the textbook answer.

It's not critical for the group to cover the topic thoroughly, and it definitely isn't important to settle on an answer. We just want everyone to have a moment to consider the question the students were working with.

The Problem with the Objective

Watching and talking about a video is a wonderful preparation and motivation for reading a case study. We find out how much of an advantage that is when we don't have a video to watch! When talking about this case study, it makes sense to review it first, to be sure everyone has some common understandings about the class it describes. A good place to start is with the problem Steve saw in the learning objective. So we generally open conversations by asking for a volunteer to give a recap: "What was the problem with MLO 4.8.3?"

In most groups at least someone would like further explanation. In part, Steve's concern came out of thinking about the subject matter itself—there's no clear line between the two kinds of change based on the properties students could observe. And Steve's concern also came out of his experience teaching the subject over the years, his sense that students ended up memorizing something they didn't really understand.

Having talked about the problem, people might have their own ideas about how to address it, either before or after talking about Steve's approach. Among other things, that gives people who've taught this material a chance to talk about what they do and maybe what their students do.

At some point the group should review how Steve decided to address the problem. Some people may have missed the flow of the activities he planned: He would first give the students a chance to come up with their own ideas, and for that he'd focus only on the "logical integrity" of their thinking. Then he would use the activity of sorting changes on the overhead projector to explain how scientists do it.

Attending and Responding to Students' Reasoning
The main agenda is to talk about interpreting and responding to student thinking. What are some things Steve saw? How did he respond? What other possibilities might there be?

Understanding the Game of Inventing Categories
The first example of Steve's interpreting and responding to student thinking came before the class discussions, in the assignment Steve gave students to categorize twelve changes they observed in lab. Their work gave him a small surprise: most of the students created categories that worked only for those twelve changes, apparently without thinking about how other situations could fit in. Their responses made him aware of something that was "second nature" to him, a tacit part of the task he realized they needed to learn: when scientists come up with categories, they want their categories to work in general, not just for a subset of whatever they're categorizing. As a first step toward helping students see that criterion for categorization, he adjusted how he presented the next assignment, making it explicit that more items could be added to the categories at any time. Another option he mentioned would have been to give a brief lecture with examples of other scientific categories.

Andy's, Lamar's, and Kathleen's Ideas
More conversations will probably start here, with the first excerpt of dialogue. The students are fairly articulate, and compared with the earlier cases, it's

relatively straightforward to understand what they are saying. Still, there can be some differences in interpretation, so it's not just going through the motions to talk about their reasoning.

For example, some people may question Steve's sense that Andy's reasoning was based entirely on the salt disappearing, pointing to the statements that "both [salt and Alka-Seltzer] change the chemical composition of the water," and that the salt "actually mixes." Maybe Andy knew that the water would now taste salty? That would still be an observable change, but a different one from the disappearance. Or maybe he was thinking the opposite, that the salt couldn't simply be gone; it must be in the water still, "actually mixed."

Steve saw Lamar's idea as different from what Andy had said, despite Lamar's claim that he was just agreeing, and Steve responded to make that clear. One person has suggested that Lamar may not really have thought he was saying the same thing, that he was getting stuck on choosing the word *matter* or *volume* or *space* and just wanted to surrender the floor. Was Lamar's idea different from Andy's or an elaboration of it?

And was it reading too much into Kathleen's contribution to call it reversibility? If she wasn't expressing the idea of reversibility, what *was* she saying? This is also a nice moment to consider the teacher's response of naming Kathleen's idea with a technical term. (Someone may remember Steve had said some students used that word in their posters, but we don't know if Kathleen was one of them.) How might that have affected the conversation? Some worry about using technical vocabulary: Does this tell students that science is terminology? Others note that here the word expressed the student's meaning; some see it as marking her idea as significant.

Skipped Opportunities

Many people want to talk about the choices Steve made not to take up the opportunities he noticed, to elaborate on the logic of Andy's reasoning or to steer the discussion toward reversibility. Later, he made a similar choice with respect to Ben's argument, which also focused on reversibility but went in the other direction, this time directly in line with the textbook answer. Steve remained neutral, as long as the ideas were logically sound, and deferred guiding the students toward the scientists' distinction until the next activity. Another option would be to use this moment to introduce the scientists' argument. What might be the advantages and disadvantages of that choice?

Helping Students Recognize Ambiguity

Later, Steve wrote about it being "a pleasure to hear the 'right' answer," after an exchange mainly with Hal and Andy. At the same time, he was concerned that they did not see the ambiguity in the idea of "making something new." Steve had planned the activity of sorting changes on the overhead projector as a way to explain where scientists draw the line. But he was still noticing and responding to the logical integrity of the students' thinking.

So when Joel argued that mixing two kinds of sand together made "a new thing," Steve saw him as having found the weakness of the definition. Steve's interpretation and choice of response might be controversial: If the students had arrived at the right answer, might it have confused them to let the discussion move away from it in this way? We might raise the question if no one else does. Wasn't Steve's purpose here to explain the scientists' categories? Similarly, some will want to talk about Steve's interaction with Betty, challenging her to make a precise distinction between the two kinds of mixing, as well as his choice in the end to give the textbook's answer and then to explain why he disagrees with it. As he described in the first section of the case study, Steve wanted the students to recognize the arbitrariness of the textbook's answer, based on what they knew, to see categorization as human invention and, often, debatable. From his perspective, students were encountering that ambiguity directly while trying to categorize the difficult case of dissolving salt.

Coordinating Multiple Objectives

As with many points that come up in talking about case studies, how people think about Steve's interpretations and choices reflects both their sense of what science is and their sense of the students' knowledge and reasoning. It also reflects the issues they are contending with in their practices.

For many, the biggest issue is time: How much can they afford to devote to activities such as these? That may come up here, although most groups will have talked about it before.

The crux of the tension for Steve was between wanting to cultivate students' sensible reasoning—its "logical integrity" and "connection to real life"—and the traditional objective of imparting correct knowledge. Here it wasn't just a matter of time; he felt this particular objective that students learn to "distinguish between chemical and physical changes based on properties" was misconceived.

In the end, his approach here was to acknowledge the conflict to the students: he told them the textbook's answer, and he told them why he found it problematic. People may want to reflect on that as a strategy, beyond the details of this case study, and that might be appropriate. It is one possibility for coordinating different objectives, when they conflict in this way. What are others?

There's another topic that may come up here: Steve's understanding of chemical and physical changes went beyond the information in the Maryland Learning Objectives. That's what let him recognize and reflect on the problem. For people with less preparation in science, it would be difficult to tell the difference between a problem with the objective and a problem with their understanding.

CHAPTER 10

Moving Forward

Watching and listening to children's inquiry in science is something like walking through a lush forest looking for food. If you know how to recognize it, there's plenty around. If you don't, if you're only used to food as it appears in the supermarket, you won't find anything. Worse yet, you might trample edible plants.

This book has been something in the direction of a field guide to the natural resources in children's inquiry, to the beginnings of science in their knowledge and reasoning. In this final chapter we start out with a quick review of what these beginnings might look like.

Then we turn to strategies for teaching. So far we've tried to avoid that topic, not because we don't think it's important but because we wanted first to establish a focus on students' thinking. We hope to have done that now, so we can think about ways to pursue that focus in class. That's our main purpose in this chapter.

Finally, we say a little about how you might collect snippets and study them for signs of nascent science in your own students.

Quick Review

"The whole of science is nothing more than a refinement of everyday thinking" about what things happen in the world and what makes them happen. We keep trying to describe that in different ways—we've called things that happen *phenomena* or *effects*; we've talked about what makes them happen in terms of *causes* and *mechanisms*. But the words are less important than the basic idea that science starts from an everyday sense of, for example

- what can make things hot, as Jessica Phelan's students showed when they wondered what was going to make the rock get hot;
- how things can move in different ways, and what can cause that, as Mary Bell's students were using to predict what would happen to the pendulum if she let it go, or as Jamie Mikeska's class was using to think about how different things fall;

- the different ways things can change, as Steve Longenecker's students were using to try to come up with categories;
- what can keep something from falling, as Kathy Swire's students were using to understand that *something* had to be happening to keep the magnet from falling; and
- what can make things take shapes, as Pat Roy's students were using to think about whether and why bubbles always come out round.

Of course there's much more. If we'd picked other cases, we'd be listing examples of children's sense of what can make things spin, or give off light, or make noise—as we said in Chapter 2, by the time children start going to school, they have an awful lot of useful knowledge about the physical world. We've settled on *sense of mechanism* as the way to refer to this knowledge, which is a term from science education research; you should use whatever term works for you.

The important point is that this knowledge is the raw material for scientific understanding. So we should help children learn to work from it, to express it in words and drawings, to refine it into something more precise and consistent.

That's the other part of everyday thinking we talked about, more beginnings of science we can see in children, in how they can find ideas in their experience and convey them to others, hear others' ideas and consider them, and, as we emphasized, detect inconsistencies and maybe even fix them. We've seen children

- finding evidence against the explanation that bubbles are round because corners would make them pop;
- finding evidence against the idea that there's magnetic metal in our bodies, and accounting for evidence that there *is* metal with another explanation;
- criticizing the consistency of arguments about whether dissolving salt is reversible or makes "something new";
- criticizing the explanation that dropping the washer on a string is like letting go of a rope swing, because one lets go at the top and the other at the bottom;
- challenging the idea that crumpling paper makes it weigh more, by arguing that it doesn't change the size of the sheet; and
- recognizing they needed an explanation of how rock gets hot, ruling out possibilities, and then choosing one that fit with what they knew.

We want to help students learn to use these abilities and refine them, too. We should help them develop tools and strategies for expressing their reasoning, for making it more consistent, for hearing and responding to others' reasoning, for keeping track of ideas people have and the evidence they collect in support or refutation. In short, we should help them use and develop their abilities for argumentation.

In sum, children have lots of valuable knowledge and abilities, and we need to recognize these resources if we're going to help children develop them. But that's not always easy to do.

Some ideas are very hard to explain, all the more so for children. Of course that's part of what we should be helping them learn, to be articulate, as Kathy was doing when she pressed her student over what exactly she meant in saying that the magnet went "through your hand." In that case, Kathy had a sense of what the child was trying to say, but in other cases it's not clear. Part of the challenge for Pat was to get a sense of what her student might have meant in saying that "the bubble comes out as, um, the square" but "it will just come out with a bubble." What did he mean?

Other times it's hard to notice scientific thinking simply because it goes by so quickly. Few people in our seminars catch the moment when Pat's student Samuel suggested an experiment with an oval bubble wand, something that would vary shape while controlling for (lack of) corners; few people catch the reason Jessica's student Lisa wanted to know whether a rock could be half metamorphic and half sedimentary.

Of course, it's inevitable that teachers will miss some of what students say, as it is inevitable in any conversation that participants don't hear or understand everything. But it's better to miss less and notice more! So . . . we practice.

Ideas for Teaching

This set of case studies has been all about providing material for teachers and student teachers to get some practice at recognizing and interpreting students' thinking. In the end, though, the point of that practice is to inform instruction. So now we'll talk a bit about what the views of science and science learning we've discussed might mean for teaching.

Framing Objectives of Science Education

The first question is *What should science lessons accomplish for children?*

Established Answers

There are two established answers to the question of what science teaching should accomplish for children. The first focuses on knowledge: At the end of a lesson, students will be able to answer particular questions correctly, or demonstrate in some other way an understanding of some fact or information. That view of the substantive objectives is deeply engrained in the educational system; it is the focus of almost all standardized tests. The second established answer is about motivation. We want children to like science, so the lesson should accomplish that, too: it should raise and maintain their interest and engagement.

For the most part, we know science teaching frequently fails most students in both of these objectives. Stop random adults in the street and ask them questions to gauge their understanding or interest in science, and you'll find most of them have little of either. We need new ways to think about science education, and not just about methods. For years educators have tried different approaches, and so the pendulum has swung back and forth between traditional methods, which tend to emphasize facts and information, and reformed methods, which tend to emphasize motivation. But it won't matter whether you stop people in their twenties or their seventies; science teaching has been failing students for a long time. Maybe it's not the methods; maybe it's the objectives.

A Different Agenda: Cultivating Resources

We've been suggesting a new way of thinking about what early science education should accomplish: cultivating the beginning of science in children's knowledge and reasoning; helping them approach science as a refinement of their everyday thinking. That's not about motivation, although we find that it does motivate children to have the opportunity to express themselves and be taken seriously. It's about the substance of their knowledge and reasoning.

But it's not about facts and information either, although eventually it will support that agenda too: children who genuinely build from what they know—who question and try to resolve contradictions, who insist that new ideas make sense—end up with much deeper understanding of the traditional facts and information than children who learn to accept things on the teacher's authority.

So the new agenda we're proposing doesn't, in the long run, conflict with the others. But it can in particular moments. Not everything children are motivated to do is helpful for science, and of course, not everything that is sensible, mechanistic, and consistent with what children know is correct.

If, for example, the children are having a wonderful time making up fanciful stories around some natural phenomenon—the gods make it thunder and rain, say—they might be motivated and engaged, and the activity might be of educational value, but it wouldn't be science. In such moments, you may need to choose between cultivating children's resources for science and letting them continue in an activity they are enjoying. If your immediate purpose is children's engagement, then they're there, and you should let them be. Jamie was in that sort of position, in "Falling Objects" (Chapter 5), when her first graders started reporting, show-and-tell style, results that couldn't make sense together.

If a child is trying to explain her reasoning for an answer you know to be incorrect, you have another choice: if your immediate purpose is correctness, you will respond differently than if it is helping the child learn to articulate her reasoning. We're suggesting that the choice in those moments should often be to focus on cultivating resources. If the particular mechanism the student explains doesn't end up leading to the right answer in this instance, it might well be part of right answers in other instances. She's exploring her mental toolbox, learning about what's there and that it's relevant for thinking in science.

Coordinating with Traditional Content

This new agenda of cultivating resources should not replace the others, certainly, but it should be coordinated with them. With respect to traditional objectives, as we discussed in Chapter 2, we expect that in general the early emphasis should be more on helping children approach science as a refinement of everyday thinking than on arriving at scientists' ideas. As they become established in that approach, the emphasis can shift to the traditional objectives.

But that's a broad, long-term view, and it may be of little help for teachers making shorter-term choices, about particular students in particular situations. In general, the challenge is greater in middle school and beyond, when curricula begin to make greater demands of coverage. Few teachers have the luxury of working with students who are so established in a meaningful approach to science that they can take it for granted, and the challenge of making traditional progress while also achieving and maintaining a meaningful approach is ongoing and often difficult.

If the students are reasoning about some topic based on what they know, and they can use what they know to understand the established idea in the objectives, then coordinating is easy. So Jessica found that her students made progress in both at the same time, once they began to start from what they knew, as their own sensible ideas were essentially correct. But her students might have gone the other way, in that moment, to treat science as a matter of memorized terminology. What if their sensible thinking took them in a different direction from the canon?

That's more in line with what Steve was trying to manage. Given what the students knew, and what the curriculum was covering, they would have to find the state learning objective about chemical and physical changes to be ambiguous and arbitrary—not good scientific practice. But he and his students would be held accountable for knowing that information. How could he help them learn the "right answer" without undermining their sense of what science is about?

These choices of objectives are all particular to the students, the subject matter, and even the school context (what if there's a high-stakes test next week?). Jessica's choice was specific to what the students had studied (igneous, sedimentary, and metamorphic rocks) and where she next needed them to go (a basic understanding of the rock cycle). Steve's was specific to the topics in his curriculum. It's beyond what we can do here to help with particular topics: it will take many books for that, each a series of case studies tied to a particular curriculum, or at least focused around particular topics. Deborah Schiffer and her colleagues are developing such materials in mathematics.

What we can do here is talk about general strategies for attending and responding to children's thinking, strategies you might integrate into your practices. We don't pretend you'll have an easy time doing that; we can only hope you'll try.

Having and Valuing Discussions

Introducing the case studies, we warned that we didn't choose them to illustrate teaching; we chose them because they show student thinking. If you were to drop in on any of the teachers' classes at random, you'd often find other activities than science talks going on. We want to be clear that these sorts of discussions are not all of science teaching.

Here we'll turn that around to say that they should be *part* of science teaching! That's our first and main suggestion for teaching methods: Value discussions about children's ideas, and value them for what they can achieve. They can help children find, use, and develop their resources for scientific inquiry. We're not thinking only of extended, full-class discussions, as have been the focus of these cases, but also of short, spontaneous exchanges, with small groups or even between two individuals.

Before we go on, we should probably say this: If you're actually interested in what the children have to say, if you're actually trying to understand them, then you're going to be having these kinds of discussions with them all the time anyway. That's pretty much this whole section in a nutshell! So you might not need to read any more about why or how to do that.

On the other hand, there are a few ways in which paying this kind of attention to children's thinking goes against the grain of what typically happens in science class, against what you and your students might expect should be taking place. So we'll say a few things about that, to make those expectations explicit and respond to them, and say a few other things as well.

Keep Track of the Purpose

The most significant conflict is over the purpose, as we've already said: the idea is to hear and help the children develop *their* ideas rather than, at least for the moment, to guide them toward scientists' ideas.

But it's one thing to decide that in advance, to know that in the abstract, and it's another to follow through in the moment. When a student says something you're not sure is right, it can be hard not to want to do something about it—to fix the idea or steer the student away from it. That objective of correctness is just so engrained that it can sneak its way in where it doesn't belong, nudging out the objective of making sense.

So look for ways to keep track of what you're trying to do. Think of students as having nuggets of insight you really want to understand—set out to discover their minds; they're the experts on what they think. If they say something that sounds bizarre, that could be the most important nugget: it's something you don't yet understand, so it's an opportunity for you to learn. If you can settle into that stance toward students—"this is where I learn about their thinking from them"—you'll have an easier time managing the urge to correct them. And they'll have an easier time understanding that you're really and truly interested in what they think.

Notice and Arrange Opportunities

In a healthy science class, opportunities for these sorts of discussions come up all the time—pretty much any time the children have things they want to ask or say about a topic in science. Unfortunately, for any of a variety of reasons, science classes don't always start out healthy in that way, so you may want to arrange opportunities for discussions. As you do, and as the children gain experience, you should find it getting easier and easier and happening more spontaneously—more and more, you'll hear the students coming up with questions and ideas that are worth discussing. That was our experience in the project.

When you do plan a discussion, and you're going to pick the question, think of one for which you expect the students might have a variety of reasonable ideas. Sometimes the variety might be different parts of the same story, such as on day one with the first graders, when they all agreed the book would fall first and contributed lots of ideas to explain why. Sometimes the variety is conflicting ideas that need to be debated and reconciled, such as in the pendulum debate when the students disagreed over three different possibilities.

That's what generally makes for the most productive discussions, a question that gets a lot of ideas out onto the table, and here's another little conflict with the usual expectations: If the purpose were to guide children to a particular idea, you'd want to ask questions that pointed to that idea, questions children would be likely get right. That's the Socratic approach of asking questions to guide students in the direction you want them to go, and sometimes that's what you need to do. But Socratic questions generally make for miserable science talks, because they don't lead to a variety of reasonable possibilities.

So, when you have a question in mind, think about it and ask yourself, "How might children answer this?" If you can think of a lot of sensible possibilities, answers a reasonable person might give, that's a good sign. If you can't, you might look for a different question.

Of course, it can be difficult to think up those possibilities. One way adults can be wrong in anticipating what children might say is in not being as imaginative as students are. In many of the snippets we saw in the project, the children invented very reasonable ideas that hadn't occurred to us in advance.* The demand on the teacher is to come up with reasonable

---

* As we mentioned in Chapter 8, we generally don't expect talks about magnets to get far, because when we think of what children might say, we don't see many options. And then Kathy Swire brought in her snippet!

possibilities for an answer, which is a different demand from that of knowing the scientists' answer.

Expect Students Have Abilities

We've heard it again and again from teachers who watch our case studies: "My students could never have a conversation like that." That, of course, can be a self-fulfilling prophesy: if the people who think this way don't give their students the opportunities, or work to make the opportunities successful, then they'll end up being right about what happens in their classes but for the wrong reasons.

We've also heard, again and again from the teachers in the project as well as from teachers and student teachers in our seminars, that they were amazed at what their students could do when they really gave them the chance. Almost any group of students can have these sorts of discussions, although this sort of discussion can take many different forms.* Part of the challenge is in recognizing what this type of discussion looks like, for any particular group, as it may be quite different depending on cultural norms, such as of appropriate tone and pace and language, or just depending on individual personalities in the room.

Another part of the challenge lies in how people may conceptualize student abilities. Traditional work in cognitive psychology described reasoning abilities as limited by children's stages of development. For more than twenty years, however, those models have been questioned and revised substantially. Developmental psychologists now understand children's abilities as far more variable and sensitive to context than those traditional accounts described. One of the clarifications of previous work is this: children are capable of abstract thinking from very young ages.

So expect that your students are capable of productive conversations. If you do, then you'll work harder and be more imaginative in trying to help. If it's not happening yet, try to figure out why. Maybe they don't know what you want them to do? Maybe they don't trust that it's really OK to say what they think? (That's why it's important for you to stay clear about your objectives, for yourself as well as for them—if you slip and start correcting them, you may send the message that they should be careful not to say anything wrong.)

Don't Think Activities Have to Be Hands-On

There's a prejudice in elementary science education that activities have to be hands-on in order to engage children. Some of that is about abilities: those old ideas about developmental stages may have contributed, with the notion that children can only think concretely. Some of that is about expectations of interest: Adults often think that conversations just can't hold children's attention.

* We say "almost" to acknowledge that there are particular exceptions, such as students with severe cognitive disabilities.

They don't expect children to be interested in *talking*; they think children need things to *do*.

It's just not true! Children love to talk, as any parent knows, especially when they think someone is listening and taking them seriously. They get excited, even passionate, about expressing themselves. Of course, science curricula should include hands-on activities. Just don't think that's the only or always the best way to engage children in a topic.

Children do, as some people have said in our workshops, "need to have experience first, to have something to talk about," but children *have had* lots of experience with the physical world, every day, all the time. For a great many questions, what they already know is more than enough for meaningful, productive discussions. Those first graders knew an awful lot about what happens to falling books and pieces of paper before they tried it in class. Those third graders knew about bubbles; the fifth and sixth graders knew about swinging.

Don't think either that it isn't scientific to reason in the absence of formal data. No competent scientists simply launch into experimentation. By the time they are actually taking data, they've spent a great deal of time discussing their ideas and expectations for what might happen, including for what data are worth collecting. For them as for students, reasoning before gathering data is a way of taking a kind of inventory of their current knowledge and understandings: What ideas can they find that relate to the topic?

In fact, for students an experiment can even get in the way of productive discussion and thought—we saw this happen on several occasions in project snippets. The purpose of scientific inquiry is to arrive at a coherent understanding, so it's always essential to consider and respond to arguments and evidence on all sides of a question. But children (and scientifically naïve adults) can think that the purpose is to arrive at *the right answer*. If they think they've seen the right answer, they might think they're done. They try an experiment, get a result, and decide that settles it without revisiting the arguments and evidence they had for other possibilities.* So sometimes it helps to hold off on the experiment.

Play a Substantive Role
Reformed science teaching is often caricatured as the teacher leaving the students to themselves. In some versions, that philosophy extends to the teacher not even asking questions. Some call it the "noble savages" approach; we think of it as the "children are natural scientists" view—just let them do their naturally scientific thing and stay out of their way. There are moments when that's a good choice, but then there are moments when it's a good choice to give answers and explanations. The hard part is in recognizing which moments those are.

---

* Naturally, that process sometimes has them settle on an incorrect result. So, for example, people who try swinging and dropping the pendulum at the end of its swing generally conclude it flies outward.

But there's another sort of role that's important to play, one that is neither leaving the students to themselves nor guiding them to discover particular answers. We're suggesting that, in early science teaching, there's a great deal to be gained by guiding children not toward particular answers, but toward kinds of thinking, the kinds of thinking we talked about in Chapter 2 and reviewed at the beginning of this chapter. Don't stay out of it when they're having these discussions. Pay close attention to what they're saying, trying to make sense of it yourself, and step in when it's appropriate to help them make progress in the direction of scientific inquiry.

Help them learn to be articulate about their ideas. Help them learn to listen, and model that yourself, such as by asking questions to clarify what they are saying. Guide them to use their sense of causes and effects, tangible ideas about what makes things happen from their everyday thinking. Help them learn to shop through their knowledge of phenomena and look for connections.

And from there, guide them to work toward consistency in their thinking. Help them learn to identify inconsistencies and try to reconcile them. Help them learn not simply to contradict someone when they don't agree, but to give evidence and reasoning and, if they can, to account for the other person's reasoning. That's what Kathy Swire's student Evan did when he explained Taylor's evidence for metal: she did see something on the x-ray, but the patient "was probably wearing something" metal.

The best part is that students *can* do these things. They don't always do them; they might not even be aware of when they're doing them and when they aren't. Much of the guidance you can give is in telling them, through your interest and attention maybe more than with actual words, "Yes! That's what you should be doing! That's science."

Keeping Science Sensible

Science lessons should and will involve many sorts of activities: everywhere from "messing about" to controlled experimentation to reading explanations in books. The most important thing is that the whole of what you do in science helps students learn science as a refinement of everyday thinking. That might not happen if the different activities send different messages, such as if in one activity children debate their ideas about how a battery makes a bulb light and in the next activity they look for vocabulary words hidden in a grid of letters (*voltage, current, amperes, power,* etc.).

So a question to ask of everything children experience in science is whether it promotes or discourages their treating science as connected to and building from what they know about the world. We talked a little in Chapter 2 about the role of terminology in this view: Children should learn technical vocabulary when and if it helps them to express and understand ideas more precisely. Learning vocabulary for its own sake might do the opposite—get in the way of their expressing their knowledge and experience or even thinking their knowledge is relevant.

Terms such as *voltage* and *current* do have connections to everyday experience, and the ideas they express can be understood with tangible meaning; everyday thinking can lead to an understanding of what makes the bulb light. Exploring those connections is beyond what we can do here, as we've discussed.

We haven't said anything, though, about science fairs, a topic that comes up all the time in our workshops. For many teachers, that's the main place to focus on children's independent thinking, through designing and conducting their own investigations. More and more schools are adopting science fairs as encouraged or required activities, with rules and strategies designed to help children learn to conduct their own investigations. The purpose is to engage children in inquiry, but the typical science-fair image of inquiry doesn't line up easily with the views we've presented here about sense of mechanisms and argumentation. We find the typical practices need some adjusting if they are to help children approach science as the refinement of everyday thinking.

Questions

The first question is about the questions students work with: How should children arrive at topics? There's usually a strong emphasis on posing questions that are *testable*, in other words, that can be answered through a manageable, controlled experiment. And so science fair materials encourage questions like Do plants grow best in red, yellow, or blue light? and Which freezes most easily: tap water, boiled water, saltwater, or sugar water? and What material catches fire most easily: newspaper, white paper, cotton balls, or science-fair poster board? Those are seen as good questions because they're testable: a student can grow plants, varying only the color of light they're exposed to; prepare samples of each kind of water and time how long each takes to freeze; hold a flame under samples of each material and time how long it takes them to ignite.

The problem is that the fact that they're testable may be the *only* reason to ask those questions. That's just not a meaningful way to approach science! And it certainly isn't a refinement of everyday thinking, in which we ask questions because we're interested for some reason in the answer.

In science, we ask questions because the answers will help us make progress in some way toward understanding what things happen in the world and what makes them happen. So we might ask about whether plants grow differently in different light *because we have some reason to suspect they might*. Our suggestion is to make sure having a reason to ask the question is part of the activity. There is simply no point to testing a question if there is no reason to ask it.

Maybe the reason to ask is that students have heard plants grow better in blue light, say, but they are suspicious because that doesn't fit with other things they know about plants or they can't think of any reason the color of light should make a difference. So, they want to check to see whether that's what happens. Or they've heard that if you boil water and then put it in the freezer, still hot, it freezes more quickly than water from the faucet. But that doesn't make sense, because in order to freeze, the hot water would have to cool down first to the temperature of the tap water. So, they want to get some evidence one way or the other.

Investigations

Typically, the investigation a student conducts for science class is a controlled experiment. That's generally the heart of the entire enterprise, with rubrics and guidance organized entirely around choosing and designing a controlled experiment. Later, advanced students may be allowed to do theoretical work; perhaps more students should have that choice.

We talked a little in Chapter 2 about how scientific practices of controlling variables are essentially forms of argumentation. A scientist controls for a variable because someone might argue it makes a difference and the scientist wants to be able to respond. That means if nobody would ever argue that some variable could make a difference, the scientist doesn't have to control for it! It's just not true in science that you never really know until you try it, because there's an infinite number of things you could try, most of which are silly.* That's how children should learn practices of controlled experiments, as parts of argumentation. So children should not put a control in their experiment because the rules say they're supposed to have a control; they should do it for some sensible reason.

Conclusions

Finally, children draw conclusions from their investigations and present their findings to others. This, too, should be about mechanisms and argumentation. They should not expect simply to report what they found—the cotton ball burned the fastest, for example. Science is about explaining what makes things happen as they do. What might it be about the cotton ball that made it burn more easily than paper? If the hypothesis going in was that paper would burn the fastest (Because it's thin and flat? Because that material is very dry?), then that's something to try to reconcile: What part of the reasoning that led to that hypothesis didn't work? All of this could even lead to a new prediction, to test one day with another experiment.

Of course, in many cases children won't be able to account for their results. That's OK; that's part of authentic science. But they should try! Maybe they can't explain their results, but they can rule out an explanation—that would be wonderful.

In other words, rather than define the process of investigations in science around testable questions and their outcomes, define it around asking

* In fact, you could have a great time with students, and it might even be productive, brainstorming things they don't have to test to find out. For example, would the boiling temperature of a pot of water in New York depend on the number of people in California who were dancing the Macarena? There's never been a study.

interesting questions, formulating ideas about their answers, and finding ways to make progress in figuring them out. The point, in sum, is that children should see and experience science as about finding ways to answer interesting, meaningful questions.

Science in Small Steps

In professional science, there are big, interesting questions, such as How old is the universe? and How did life begin? Those would be pretty hard for children to explore in science fair projects, of course, but they're the sorts of questions that rouse passions and drive scientists throughout their careers. Thinking about them leads to other smaller questions that might help. Some scientists, because they want to know how old the universe is, have worked on figuring out the processes of how stars form and die, and that in turn has connected to more specific questions they could study in the lab about how particular bits of matter interact. The point is that children should understand that questions like How old is the universe? are most definitely asked and valued in science.

As are more kid-sized questions, such as What causes lightning? and Why do eggs get hard when you cook them? Then the challenge, as in professional science, is to find ways to make progress. Of course, students could look up the answers that scientists have found in the encyclopedia, but there's no point to that if the idea is for them to conduct their own investigations.

Maybe a child has heard that lightning is caused by static electricity, and she wants to try to confirm that. Can something so big and powerful really be the same as something so tiny? She finds a way to make static electric sparks, maybe building a device from a design she finds on the Internet, to see if they look like lightning. The sparks are very tiny, so she gets some help at magnifying them, or maybe at making bigger sparks, and borrows a video camera to get pictures. She sees that they do look like little lightning bolts and puts pictures showing that in her project, and that finding adds credence to her idea that lightning is static electricity. She hasn't answered her original question yet (What causes lightning?), but she's made good progress: she's decided it might be the same thing that causes little sparks. (Don't take points off because there's no controlled variable! She's done nice work, and not all of science involves controls!)

Or maybe a child thinks it's weird that heating eggs makes them hard. Heating so many other things makes them melt! Maybe it's the same reason that mud gets hard: cooking an egg dries it out. How could he test that idea? If the egg is drying out, it should weigh less, so maybe he could find the weight of a raw egg and compare that with the weight of the egg after cooking. He'd want to repeat that a few times, so he'd know he didn't just have a weird egg, and keep careful track of his data, and be sure the scale didn't get wet because that would throw off his measurements. If he thinks someone might argue that the temperature of the egg would make a difference—a classmate could think hot eggs weigh less because heat rises—then he should control for temperature. And maybe he'll want to use mud for comparison: How much does the weight of mud change when it dries? He, too, might not end his project with an answer to his original question, but he might rule out his idea about drying, and that would be terrific.

What's Next?

At the beginning of this chapter we wrote that this book has been "something in the direction of a field guide to children's inquiry." It's only in the direction of that: We don't yet know as much about cognitive resources as, say, botanists and woodspeople know about edible plants. But we're making good progress: we know to look for children's resources for understanding physical causes and effects (mechanisms) and for understanding and reconciling different points of view (argumentation). But like wild mushrooms, children's ideas come in many varieties and have many appearances. It's impossible to provide a catalog of resources and their various manifestations in what children say and do. Recognizing them in any particular student's thinking is something that, in the end, you have to learn to do for yourself.

Looking forward from here, we hope to have helped you notice, appreciate, and elicit the substance of your students' knowledge and reasoning. If you're new to science—we've written these materials expecting that you are—we also hope to have helped you notice and appreciate the substance of your own thinking. That's what we found in our project. It wasn't just that studying physics during the summers helped teachers work with their students over the year; it was also the other way around, and maybe more so. All the time we'd spent over the year talking about what we could hear and understand in children's inquiry seemed to give the teachers a different view of how they should approach science themselves, and the project staff were amazed by what the teachers accomplished.

If you're moved to study more science, that's wonderful. You'll want to be selective, though: Not all science courses will be helpful! Stay away from any that will ask you to accept things that don't make sense. Try instead to find a course that treats science as a refinement of everyday thinking—look in the syllabus for descriptions or hints of the course philosophy, talk to the professor about her or his approach, talk to some former students. The course shouldn't cover too many topics: it is much more valuable to experience science in a rich, meaningful way, and that takes time.

Collecting and Discussing Your Own Snippets

Our fondest hope, of course, is that you'll go collect snippets of your own and get together with your colleagues to discuss (among other things surely!) the substance of your students' thinking. All you really need are some colleagues—even if you find only one other person who wants to participate, it can be productive. Then you can collect examples of your students' thinking, on video as we've done here, or in copies of written work, or even in your detailed notes from class.

If you have a video camera, or if there's one at your school, it's easy enough to set it somewhere in the room, point it in a likely direction, and turn it on. Tape often! That way you'll have lots of classes to choose from to show your colleagues—it's hard to know beforehand what a class is going to be like and whether you'd like to discuss it with others. Collecting video of your classes could also serve you in other ways, such as providing material you'd need to apply for National Board Certification.

For most purposes, the microphone that comes with the camera will be enough to capture students' conversation; that's what you heard in Jamie Mikeska's first day's clip of the first graders. But you could try using external microphones, which you can place among the students to get better sound; we've used Crown Sound Grabbers, but there are many other options. We've also sometimes used radio microphones on the teacher's lapel, which is why Mary Bell's voice is so much easier to hear than the students'. Again there are lots of options, but we used ones made by Azden.

Don't worry if there's nobody to stand at the camera. Just place it somewhere and point it in a direction that's likely to have some action. If a colleague wants to help and tape you, all the better. Ask him or her to keep it simple: don't zoom in and out or pan around too much. This isn't cinematography; you're just trying to make a record of what happened. If you're not going to make the videotapes public—that is, if they're for your personal use only—you can skip most of the hassle we went through of getting permission forms signed by parents and guardians. Of course, your school may have its own policies.

If you can't or don't want to videotape, you could get a lot out of audiotaping. And you don't even need to do that. If after a class discussion, you take some time to sit down and write everything you remember about what students said, trying to reconstruct dialogue so as to have something like a transcript, that can give you and colleagues something to review and consider. You'll get some things wrong in remembering, but that's OK. The truth is that any way you keep a record of what happens in class will miss some things, including video.

But you'll want to have some way of focusing careful attention on particular things children say, write, and do, and trying to understand their thinking. That's the point of all this, don't forget—and try to remember that when you're talking with colleagues. Don't talk just about the teacher! Take a look at the prompt sheet we used for our conversations in the project, which we've provided on the DVD-ROM, and maybe use it to help you get started.

So give it a try! You don't have to write up a formal case study; the teachers in our project had many wonderful, productive conversations about snippets they never used for case studies. But maybe you'll be moved to as you go, and that would be great. Maybe you'll even feel like making your work public!

Notes

Chapter 2

13 **children as natural scientists** A number of authors have developed careful, insightful, and sometimes surprising accounts of the ways in which children act like scientists. Gopnik, Meltzoff, and Kuhl (1999) give a nice overview.

14 **Misconceptions** Research on misconceptions really started to take off in the 1980s (Carey 1986; Champagne, Klopfer, and Anderson 1980; Clement 1982, 1983; McCloskey 1983; Strike and Posner 1985).

15 **research didn't do a good job...** There are exceptions to this, including research by Clement and his colleagues to show that some preconceptions are useful (Clement, Brown, and Zeitsman 1989).

15 **one famous misconception** This misconception was made famous by a documentary called *A Private Universe* (Sadler, Schneps, and Woll 1989) that aired on public television and is often shown in science education programs. It opens at a Harvard graduation, with brief interviews of graduates who explain that it is hotter in the summer than in the winter because the earth gets closer to the sun.

17 **The Many Parts of Common Sense** This perspective of common sense as made up of many smaller parts has been formulated in various versions by research in cognitive science, developmental psychology, and science education (Dennett 1991; diSessa 2000; Karmiloff-Smith 1992; Lakoff 1987; Minsky 1986; Minstrell 1992; Rumelhart 1980).

18 **We call those parts resources** The original idea of schemata, which dates back to Bartlett (1932), gave a similar view. We don't use the word *schemata*, though, because it's so often part of stage-based theories of development. Other terms for related ideas have included *facets* (Minstrell 1992), *agents* (Minsky 1986), and *p-prims* (diSessa 2000).

19 From the children's perspective   Most young children will not know that the seasons are different in different hemispheres. In fact, children do not generally think about the earth as a solid sphere. Earlier work described students as reasoning consistently from one "mental model" of the earth (Vosniadou and Brewer 1992); recent research has shifted toward evidence of greater variability in students' reasoning (Vosniadou, Skopeliti, and Ikospentaki 2004).

20 the history of science   See Westfall (1978) for an account of the transition in Western thought from "naturalism" to the "mechanical philosophy."

21 Common sense of mechanisms   The importance of mechanism has been discussed in science education research (diSessa 1993; Newton and Newton 2000; White 1993) and in studies of professional science, both modern (Machamer, Darden, and Craver 2000) and historical (Westfall 1978).

21 Shopping for Ideas   The importance of analogical reasoning in science has been discussed from various research perspectives (Clement 1991; Dunbar 1995; Harré 1970; Hesse 1966).

22 cases that aren't in the book   These are all examples from snippets we discussed in the project, by Alison Alevy (fifth graders on solar water heaters), Jamie Mikeska (first graders on seeds), Patricia Roy (third graders on earthquakes), and Charles Gale (fourth graders on lightbulbs). The earthquakes snippet is the subject of a research article (May, Hammer, and Roy under review); see also Hammer (2004).

23 Reconciling Inconsistencies: Arguments and Counter-Arguments   The importance of attending to students' abilities to engage in argumentation has been a theme in science education research (Driver, Newton, and Osborne 2000; Duschl 1990; Kuhn 1991).

25 lots of examples in our project   These examples come from snippets by Cynthia Cicmansky (third graders on water cooling), Alison Alevy (fifth graders on solar ovens), Kathleen Hogan (pre-K children on day and night), and Trisha Kagey (third graders on rainbows).

28 From Footholds to Principles   For an interesting account of this sort of commitment to principles, see Feynman (1965). Research on the principled coherence of expert understanding has been a theme in physics education research (Chi, Feltovich, and Glaser 1981; Eylon and Reif 1984; Larkin 1983).

28 epicycles   Westfall (1978) begins with a brief account of epicycles.

31 the distance moved divided by the time   We're cheating a little here, but only a little. Physicists define the velocity of an object as the change in its position divided by the time interval *in the limit of a very small time interval*. That way, if the velocity is changing a lot, the time interval is small enough to think about how fast the object is moving at that instant. This is a perfect example of something everyone could learn, starting from everyday thinking, but it takes some time and attention.

36 **Of course they will**   See the lovely essay "Messing About in Science" in Hawkins (1974).

36 **Most have learned to separate what they learn in science from common sense**   There is a sizable body of research into students' beliefs about what sorts of knowledge and reasoning are appropriate in science. The findings are consistent across a range of studies that older students do not generally understand how to learn science or how scientists arrive at their conclusions (Carey et al. 1989; Duschl 1990; Hammer 1994; Hewson 1985; Ryder, Leach, and Driver 1999; Sandoval and Morrison 2003). Meanwhile, studies of young children show them to be capable of sophisticated thinking about knowledge and reasoning in science (Herrenkohl et al. 1999; Samarapungavan 1992; Smith et al. 2000; Tytler and Peterson 2003).

Chapter 3

41 **as Joseph McDonald (1992) put it**   "Teaching, closely read, is messy: full of conflict, fragmentations, and ambivalence. These conditions of uncertainty present a problem in a culture that tends to regard conflict as distasteful and that prizes unity, predictability, rational decisiveness, certainty. This is a setup: Teaching involves a lot of 'bad' stuff, yet teachers are expected to be 'good.'" (McDonald 1992, p. 21).

41 **Draws Attention Away from the Students**   Sherin and Han (2004) discuss teachers' progress in a video club from focusing on the teachers' methods to focusing first on the students' thinking. Nelson and Sassi (2000) discuss a similar shift as important for supervision.

41 **as long as the data included evidence of those ideas**   For example, *What Children Bring to Light* (Shapiro 1994) focuses on children's understandings of light and highlights how, in some classes, the teachers do not recognize or engage those understandings.

42 **there are many excellent books and websites**   Almost any book on elementary science teaching includes ideas for methods and materials. See, for example, the Science Snackbook series, published by the Exploratorium in San Francisco, including activities available on the Web: www.exploratorium.edu/snacks/index.html.

Chapter 5

71 **First Graders Discuss Dropping a Book and a Piece of Paper**   For more on the falling objects discussion, see Hammer, Russ, Mikeska, and Scherr (in press).

91   (footnote) See the chapter notes for a little more explanation   It's a surprise for many people that a brick, say, falls to the ground in the same amount of time as a penny. They're in good company: Aristotle thought the brick would fall faster, too. It's a very reasonable thing to think, because the brick is so much *heavier*—it's pulled to the ground much harder than the penny. That's a nice, mechanistic reason.

There's another part of common sense about bricks and pennies that's relevant, too. Imagine a brick is hanging from a string, and you come along and flick it with your finger to get it moving. If you're imagining that, you're probably thinking of it hurting your finger, and you're probably not thinking of the brick flying off in any hurry. That intuition is very useful: It's hard to get the brick moving! If it were a penny hanging from the string, you could probably get that moving pretty well, flicking it with your finger.

But if it's harder to get the brick moving, then shouldn't it fall more slowly than the penny? This is how physicists explain why a brick and a penny hit the ground at the same time: the brick is much harder to get moving, but it's being pulled much harder toward the ground. So the physicists' explanation puts two different parts of common sense together and uses them at the same time.

And again, as the children were discussing, for a piece of paper, the air makes a big difference. If either the brick or the penny were affected more by air—say if you dropped them from a tower so they got moving more quickly—then the results would change. That's still another part of common sense—what air can do to things—to add in to the mix for a better understanding.

92   moving away from stage-based models   For an overview of how cognitive psychology is moving away from stage-based models, see Siegler (1996) or Feldman (1994).[Feldman, D. H. (1994). *Beyond Universals in Cognitive Development*. Norwood, N.J.: Ablex.]

Chapter 6

96   Eighth Graders Discuss the Rock Cycle   For more on the rock cycle conversation, see Rosenberg, Hammer, and Phelan (2006).

Chapter 7

126   Why the Bubbles Come Out Round?   One way to understand why bubbles come out spherical is this: The material of the bubble is the soap film, which is like a balloon sheet except that it's completely fluid. A balloon has a shape to it because although the rubber of the balloon is stretchy, it's still solid. Soap film is fluid—it doesn't have any definite shape.

So why does it make a sphere? Like an inflated balloon, the soap film is always pulling inward—this is that idea of surface tension—trying to get smaller. Any sheet of soap film, pulling in on itself and able to take whatever shape results, will be as small as it can be. A typical bubble is a sheet of soap film wrapped around some air, and the sheet pulls in on itself to be as small as possible. The air inside limits how much the sheet can shrink, unless the sheet breaks or tears, and then it can shrink down to a drop of soapy water.

It has the shape of a sphere because that's the shape with the smallest possible area, holding the amount of air inside. That's the soap film making itself as small as it can be. If there were a corner, the soap film would pull it out, like when you pull out the wrinkles of a bed sheet. For it to keep that corner, something would have to help it hold that shape against the pull.

Chapter 8

144 **Physicists call that idea "action at a distance"** Can two objects have an effect on each other from a distance, without anything passing between them? Scientists wrestled extensively over how to think about magnets, which they found very hard to explain in mechanistic terms. Two objects affecting each other from a distance was also part of Newton's theory of gravity, that the earth pulls on things without touching them. But he was very uncomfortable with it. In fact, he described the idea of action at a distance as "so great an absurdity" that no one with a "competent faculty of thinking" could believe it. He complained that he could figure out how strong the force between two objects was, but had no explanation for what caused it.

That's part of why physicists came to talk about the idea of a field, as the next step in refining their understanding. So we speak of a *gravitational field* and a *magnetic field*, and we think of it as traveling between the two objects. In that way of thinking, magnetic fields do pass through your hand, and the reason they don't hurt is that they have much too small an effect on the matter in your hand to do anything—just as, say, sound can pass through your hand without hurting it. Thinking about fields is a refinement that leads to lots of further refinements, including new ideas about light (electromagnetic waves), and for more on that, you should find a good physics course!

Chapter 10

166 **materials in mathematics** Our project was especially influenced by the work of Deborah Schifter and her colleagues (Schifter 1996a, 1996b, 1996c; Schifter and Fosnot 1993). *What's Happening in Math Class* (1996b) has many case studies that cut across topics in mathematics to focus on general aspects of what it means to pursue reformed practices. In later work, these researchers have gone on to develop curriculum materials (Schifter, Bastable, and Russell 1999), as well as a series of casebooks tied to particular topics (e.g., Bastable et al. 2002; Schifter et al. 2002).

169   **children are capable of abstract thinking**   Siegler (1996) is a general starting point for reading about progress in cognitive psychology past stage-based models of development. Metz (1995) focuses specifically on how stage-based models of development have underestimated children's abilities in science.

170   **Children love to talk**   We thank Gretchen Walker, of the American Museum of Natural History, for raising the topic of educators' prejudice for hands-on activities. Her experience in after-school science programs gave some evidence of children's interest in simply having time to talk. Asked to say what they liked most about the programs, almost all the children said it was the discussions that let them express their own ideas and hear each other's. See Walker, Wahl, and Rivas (2005).

174   **interesting, meaningful questions**   Some schools have changed the traditional science fair into a "science inquiry conference," in which children discuss their questions and findings with one another. See Saul et al. (2005).

175   **what the teachers accomplished**   For a case study of work that Mary, Pat, and another project teacher, Jennifer Peter, did with studying light, see van Zee et al. (2005).

175   **You'll want to be selective**   There have been a number of textbooks and materials developed for teaching physics in more meaningful ways, including courses specifically for elementary teachers (American Institute of Physics and American Association of Physics Teachers 1995; Laws 1997; McDermott 1996). Of course, much depends on how the particular course uses these materials. A brand-new course we recommend is called *Physics for Elementary Teachers* (Goldberg, Otero, and Robinson, 2006). It combines activities of reasoning in physics with activities of watching and listening to children's reasoning on the same topics.

References

American Institute of Physics and American Association of Physics Teachers. 1995. *Powerful Ideas in Physical Science: A Model Course.* College Park, MD: American Association of Physics Teachers.

Bartlett, F. C. 1932. *Remembering: A Study in Experimental and Social Psychology.* Cambridge, UK: University Press.

Bastable, V., D. Schifter, S. J. Russell, J. B. Lester, and Teaching to the Big Ideas (Project). 2002. *Examining Features of Shape: Facilitator's Guide: Geometry.* Parsippany, NJ: Dale Seymour.

Carey, S. 1986. "Cognitive Science and Science Education." *American Psychologist* 41 (10): 1123–30.

Carey, S., R. Evans, M. Honda, E. Jay, and C. Unger. 1989. "'An Experiment Is When You Try It and See if It Works': A Study of Grade 7 Students' Understanding of the Construction of Scientific Knowledge." In *International Journal of Science Education*, vol. 11, ed. R. Driver, 514–29.

Champagne, A. B., L. E. Klopfer, and J. H. Anderson. 1980. "Factors Influencing the Learning of Classical Mechanics." *American Journal of Physics* 48: 1074–79.

Chi, M. T. H., P. J. Feltovich, and R. Glaser. 1981. "Categorization and Representation of Physics Problems by Experts and Novices." *Cognitive Science* 5: 121–52.

Clement, J. 1982. "Student Preconceptions in Introductory Mechanics." *American Journal of Physics* 50: 66.

———. 1983. "A Conceptual Model Discussed by Galileo and Used Intuitively by Physics Students." In *Mental Models*, ed. D. Gentner and A. Stevens, 325–40. Hillsdale, NJ: Lawrence Erlbaum.

———. 1991. "Nonformal Reasoning in Experts and in Science Students: The Use of Analogies, Extreme Cases, and Physical Intuition." In *Informal Reasoning and Education*, ed. J. Voss, D. Perkins, and J. Siegel, 345–62. Hillsdale, NJ: Lawrence Erlbaum.

Clement, J., D. Brown, and A. Zeitsman. 1989. "Not All Preconceptions Are Misconceptions: Finding 'Anchoring Conceptions' for Grounding Instruction on Students' Intuitions." *International Journal of Science Education* 11: 554–65.

Dennett, D. C. 1991. *Consciousness Explained.* Boston: Little, Brown.

diSessa, A. 1993. "Towards an Epistemology of Physics." *Cognition and Instruction* 10 (2–3): 105–225.

———. 2000. *Changing Minds: Computers, Learning, and Literacy.* Cambridge, MA: MIT Press.

Driver, R., P. Newton, and J. Osborne. 2000. "Establishing the Norms of Scientific Argumentation in Classrooms." *Science Education* 84 (3): 287–312.

Dunbar, K. 1995. "How Scientists Really Reason: Scientific Reasoning in Real-World Laboratories." In *Mechanisms of Thought*, ed. R. J. Sternberg and J. Davidson, 365–95. Cambridge, MA: MIT Press.

Duschl, R. A. 1990. *Restructuring Science Education.* New York: Teachers College Press.

Einstein, A. 1936. "Physics and Reality." *Journal of the Franklin Institute* 221 (3): 349–82.

Eylon, B.-S., and F. Reif. 1984. "Effects of Knowledge Organization on Task Performance." *Cognition and Instruction* 1: 5–44.

Feldman, D. H. 1994. *Beyond Universals in Cognitive Development.* Norwood, NJ: Ablex.

Feynman, R. P. 1965. *The Character of Physical Law.* Cambridge, MA: MIT Press.

Gallas, Karen. 1995. *Talking Their Way into Science: Hearing Children's Questions and Theories, Responding with Curricula.* New York: Teachers College Press.

Goldberg, F., V. Otero, and S. Robinson. (2006). *Physics for Elementary Teachers.* Armonk, NY: It's About Time.

Gopnik, A., A. N. Meltzoff, and P. K. Kuhl. 1999. *The Scientist in the Crib: What Early Learning Tells Us About the Mind.* New York: HarperCollins.

Hammer, D. 1994. "Students' Beliefs About Conceptual Knowledge in Introductory Physics." *International Journal of Science Education* 16 (4): 385–403.

———. 2004. "The Variability of Student Reasoning, Lecture 1: Case Studies of Children's Inquiries." In *Proceedings of the Enrico Fermi Summer School, Course CLVI*, ed. E. Redish and M. Vicentini, 279–99. Bologna, Italy: Italian Physical Society.

Hammer, D., Russ, R., Scherr, R. E., and Mikeska, J. (to appear). Identifying inquiry and conceptualizing students' abilities. In R. A. D. a. R. E. Grandy (Ed.), *Reconsidering the Character and Role of Inquiry in School Science: Framing the Debates.* Rotterdam: Sense Publishers.

Harré, R. 1970. *The Principles of Scientific Thinking.* Chicago: University of Chicago Press.

Hawkins, D. 1974. *The Informed Vision: Essays on Learning and Human Nature.* New York: Agathon.

Herrenkohl, L. R., A. S. Palincsar, L. S. DeWater, and K. Kawasaki. 1999. "Developing Scientific Communities in Classrooms: A Sociocognitive Approach." *Journal of the Learning Sciences* 8 (3–4): 451–93.

Hesse, M. B. 1966. *Models and Analogies in Science.* Notre Dame, IN: University of Notre Dame Press.

Hewson, P. W. 1985. "Epistemological Commitments in the Learning of Science: Examples from Dynamics." *European Journal of Science Education* 7 (2): 163–72.

Karmiloff-Smith, A. 1992. *Beyond Modularity*. Cambridge, MA: MIT Press.

Kuhn, D. 1991. *The Skills of Argument*. Cambridge: Cambridge University Press.

Lakoff, G. 1987. *Women, Fire, and Dangerous Things: What Categories Reveal About the Mind*. Chicago: University of Chicago Press.

Larkin, J. 1983. "The Role of Problem Representation in Physics." In *Mental Models*, ed. D. Gentner and A. Stevens, 75–98. Hillsdale, NJ: Lawrence Erlbaum.

Laws, P. W. 1997. *Workshop Physics Activity Guide*. New York: Wiley.

Machamer, P., L. Darden, and C. F. Craver. 2000. "Thinking About Mechanisms." *Philosophy of Science* 67 (1): 1–25.

Maryland County Public Schools (MCPS). *Curriculum Framework*. 2001.

May, D. B., D. Hammer, and P. Roy. Under review. Children's Analogical Reasoning in a 3rd-Grade Science Discussion.

McCloskey, M. 1983. "Naïve Theories of Motion." In *Mental Models*, ed. D. Gentner and A. Stevens, 299–324. Hillsdale, NJ: Lawrence Erlbaum.

McDermott, L. C. 1996. *Physics by Inquiry*. New York: Wiley.

McDonald, J. P. 1992. *Teaching: Making Sense of an Uncertain Craft*. New York: Teachers College Press.

Metz, K. E. 1995. "Reassessment of Developmental Constraints on Children's Science Instruction." *Review of Educational Research* 65 (2): 93–127.

Minsky, M. L. 1986. *Society of Mind*. New York: Simon and Schuster.

Minstrell, J. 1992. "Facets of Students' Knowledge and Relevant Instruction." In *Research in Physics Learning: Theoretical Issues and Empirical Studies*, ed. R. Duit, F. Goldberg, and H. Niedderer, 110–28. Kiel, Germany: IPN.

Nelson, B. S., and A. Sassi. 2000. "Shifting Approaches to Supervision: The Case of Mathematics Supervision." *Educational Administration Quarterly* 36 (4): 553–84.

Newton, D. P., and L. D. Newton. 2000. "Do Teachers Support Causal Understanding Through Their Discourse When Teaching Primary Science?" *British Educational Research Journal* 26 (5): 599–613.

Rosenberg, S. A., Hammer, D., and Phelan, J. Forthcoming. "Multiple Epistemological Coherences in an Eighth-Grade Discussion of the Rock Cycle." *Journal of the Learning Sciences*.

Rumelhart, D. E. 1980. "Schemata: The Building Blocks of Cognition." In *Theoretical Issues in Reading and Comprehension*, ed. R. J. Spiro, B. Bruce, and W. F. Brewer, 38-58. Hillsdale, NJ: Lawrence Erlbaum.

Ryder, J., J. Leach, and R. Driver. 1999. "Undergraduate Science Students' Images of Science." *Journal of Research in Science Teaching* 36 (2): 201–19.

Sadler, P. M., M. H. Schneps, and S. Woll. 1989. *A Private Universe*. Santa Monica, CA: Pyramid Film and Video.

Samarapungavan, A. 1992. "Children's Judgments in Theory Choice Tasks: Scientific Rationality in Childhood." *Cognition* 45: 1–32.

Sandoval, W. A., and K. Morrison. 2003. "High School Students' Ideas About Theories and Theory Change After a Biological Inquiry Unit." *Journal of Research in Science Teaching* 40 (4): 369–92.

Saul, W., D. Dieckman, C. Pearce, and D. Neutze, with M. Dieckman, H. Buck, and L. Green. (2005). *Beyond the Science Fair: Creating a Kids' Inquiry Conference*. Portsmouth, NH: Heinemann.

Schifter, D. 1996a. "A Constructivist Perspective on Teaching and Learning Mathematics." *Phi Delta Kappan* 77 (7): 492–99.

———, ed. 1996b. *What's Happening in Math Class? Volume 1: Envisioning New Practices Through Teacher Narratives.* New York: Teachers College Press.

———, ed. 1996c. *What's Happening in Math Class? Volume 2: Reconstructing Professional Identities.* New York: Teachers College Press.

Schifter, D., and C. T. Fosnot. 1993. *Reconstructing Mathematics Education: Stories of Teachers Meeting the Challenge of Reform.* New York: Teachers College Press.

Schifter, D., V. Bastable, and S. J. Russell. 1999. *Developing Mathematical Ideas.* Parsippany, NJ: Dale Seymour.

Schifter, D., V. Bastable, S. J. Russell, and Teaching to the Big Ideas (Project). 2002. *Examining Features of Shape: Casebook: Geometry.* Parsippany, NJ: Dale Seymour.

Shapiro, B. 1994. *What Children Bring to Light.* New York: Teachers College Press.

Sherin, M. G., and S. Y. Han. 2004. "Teacher Learning in the Context of a Video Club." *Teaching and Teacher Education* 20 (2): 163–83.

Siegler, R. S. 1996. *Emerging Minds: The Process of Change in Children's Thinking.* New York: Oxford University Press.

Smith, C. L., D. Maclin, C. Houghton, and M. G. Hennessey. 2000. "Sixth-Grade Students' Epistemologies of Science: The Impact of School Science Experiences on Epistemological Development." *Cognition and Instruction* 18 (3): 349–422.

Strike, K. A., and G. J. Posner. 1985. "A Conceptual Change View of Learning and Understanding." In *Cognitive Structure and Conceptual Change*, ed. L. H. T. West and A. L. Pines, 211–31. New York: Academic.

Thomson, W., Sir (Lord Kelvin). 1889. *Electrical Units of Measurement. Popular Lectures and Addresses.* , Vol 1. London: Macmillan.

Tytler, R., and S. Peterson. 2003. "Tracing Young Children's Scientific Reasoning." *Research in Science Education* 33 (4): 433–65.

van Zee, E. H., D. Hammer, M. Bell, P. Roy, and J. Peter. (2005) "Learning and Teaching Science as Inquiry: A Case Study of Elementary School Teachers' Investigations of Light." *Science Education* 89(6): 1007–1042.

Vosniadou, S., and W. F. Brewer. 1992. "Mental Models of the Earth: A Study of Conceptual Change in Childhood." *Cognitive Psychology* 24 (4): 535–85.

Vosniadou, S., I. Skopeliti, and K. Ikospentaki. 2004. "Modes of Knowing and Ways of Reasoning in Elementary Astronomy." *Cognitive Development* 19 (2): 203–22.

Walker, G., E. Wahl, and L. Rivas. 2005. *NASA and Afterschool Programs: Connecting to the Future.* Washington, DC: NASA.

Westfall, R. S. 1978. *The Construction of Modern Science: Mechanisms and Mechanics.* Cambridge: Cambridge University Press.

White, B. 1993. "Intermediate Causal Models: A Missing Link for Successful Science Instruction." In *Advances in Instructional Psychology*, vol. 4, ed. R. Glaser, 177–252. Hillsdale, NJ: Lawrence Erlbaum.